CAMBRIDGE

Higher

MATHEMATICS
GCSE for AQA
Problem-solving Book

Tabitha Steel, Coral Thomas, Mark Dawes and Steven Watson

CAMBRIDGE
UNIVERSITY PRESS

University Printing House, Cambridge CB2 8BS, United Kingdom

Cambridge University Press is part of the University of Cambridge.

It furthers the University's mission by disseminating knowledge in the pursuit of education, learning and research at the highest international levels of excellence.

www.cambridge.org
Information on this title:
www.cambridge.org/ukschools/9781107450073 (Paperback)

First published 2015

Printed in the United Kingdom by Latimer Trend

A catalogue record for this publication is available from the British Library

ISBN 978-1-107-45007-3 Paperback

Additional resources for this publication at www.cambridge.org/ukschools

Cover image © 2013 Fabian Oefner www.fabianoefner.com

Contents

Introduction   iv

Problem-solving strategies:

1   Draw a diagram   1
2   I can't do that but I can do this   15
3   Adding lines   29
4   Solve a simpler problem   41
5   Make changes to the problem   51
6   Do you recognise the problem?   60
7   What is a sensible answer?   72
8   It depends how you look at it   82
9   Work systematically   96
10 If you don't know what to do, do something   103

Worked solutions:

1   Draw a diagram   112
2   I can't do that but I can do this   127
3   Adding lines   139
4   Solve a simpler problem   153
5   Make changes to the problem   161
6   Do you recognise the problem?   171
7   What is a sensible answer?   182
8   It depends how you look at it   190
9   Work systematically   206
10 If you don't know what to do, do something   214

Introduction

What is a mathematical problem?

In everyday speech you might talk about a 'problem' as being something that is negative and often annoying. It might also be referred to as a difficulty, or as something that needs to be fixed.

Mathematical problems might be difficult, they might even be annoying, but they are certainly not negative! Some people see mathematical problems as any sort of mathematical question and some people use 'problem' in this way. For some people, 2 + 3 is just an easy question. 117 × 495 is more difficult but it is still just a question (even though some people will not know how to work it out).

The definition of a mathematical problem that this books uses is: something you don't know how to solve immediately.

117 × 495 could be worked out by writing down the number 117, adding on 117, and adding on another 117 again and again until you have written it 495 times. Even if you don't have a quicker way of working this out (and it would be a good idea to have a quicker way!) then this is not actually mathematical problem solving.

You often need to think 'around' a problem before you can get started.

A mathematical problem, then, is likely to be challenging, and it might well be the case that you can't 'see' the answer quickly. There are lots of things that might help you though, and this book will show you some of those ideas.

Different types of mathematics

To be able to do maths you need to know and understand mathematical ideas. These are important, and Cambridge University Press's *GCSE Mathematics for AQA Higher Student Book* covers this sort of material. You also need to be able to solve problems, and this can be more difficult.

Problem-solving 'muscle memory'

When you learn to do something new you usually have to think very hard about what you are doing. For example, when someone first learns to drive a car they have to concentrate hard and decide which foot to press down, which way to move the gear stick, which mirror to look in, and so on. Experienced motorists still need to concentrate on the road, but they don't need to think about the mechanics of every single action they take.

This is known as 'muscle memory'. Muscle memory comes up lots in sports, so a footballer might have to train hard to learn how to do a rainbow flick, but will later be able to do it without thinking about it.

When you solve an equation the answer itself is often not very exciting ("$x = 3$ – big deal!"). The important thing when solving lots of equations is not finding out whether x happens to be 3 or 4 this time around, but that you are putting equation-solving in your muscle memory. Some of the problems you solve in this book *will* be interesting in their own right. We hope you will enjoy the challenge of doing them and will enjoy the answers too. Beyond this, though, you should not just work out the answers and be satisfied with that. After each problem, try to reflect on how you solved it. It is important to think about this, because this is the point at which you will learn new skills and this is how the muscle memory will start to develop.

Impossible to obvious

At first, a mathematical problem may appear to be impossible. When you know the answer or when you know how to do it, the problem can then seem obvious and not worth thinking about.

Hopefully you will avoid both of these extremes. None of the problems you will be asked to solve in this book are impossible – see them as a challenge. Even if a problem is obvious to you then there are still likely to be important ideas that you might be able to apply to other, more challenging problems in the future.

> A song by the band *Busted* has the chorus lyric: 'I've been to the year 3000, … and your great, great, great-granddaughter is pretty fine.'
>
> You might know the song (*Year 3000*) and might be concerned about the number of times 'great' is in the lyrics. First of all, how do we know we need more 'great's?
>
> Secondly, how many 'great's should there be?
>
> - Start working on this problem by yourself.
> - Talk to other people about what you have done.

Start working on this problem by yourself. Once you've finished, talk to other people about what you have done.

How to use this book

Each chapter in this book covers a problem-solving strategy – a way of approaching a problem. The chapter starts by describing the strategy and how to apply it to a problem. This is followed by a number of questions that you can practise applying the strategy to.

Each question covers one or more strands of maths, which is shown by the colour wheel beside the question number.

The exploded piece(s) of the colour wheel tells you what strand(s) of maths each question covers:

Number

Algebra

Ratio, proportion and rates of change

Geometry and measures

Probability

Statistics

The questions have been written to give you regular practice at problem solving so that you can build your skills, confidence and mathematical experience not only for your GCSE exams but also for life outside the classroom.

The star rating suggests the amount of problem-solving experience needed to tackle each question.

⭐ ⭐ ⭐ 1 star questions are 'entry-level'; they are good questions to start with.

⭐ ⭐ ⭐ 2 star questions are ideal to move on to when you have confidently and successfully completed some of the entry-level questions.

⭐ ⭐ ⭐ 3 star questions are for when you need more of a challenge.

In other words, the more stars, the more difficult the question is.

This means that you might need a calculator to work through the problem.

This means that you should work through a question without a calculator. If this is not present, you can use a calculator if you need to.

Tip

Tip boxes provide hints to help you work through questions.

Worked solutions to all of the questions are provided in the back of the book. Blue boxes alongside the solutions guide you through the working.

Draw a diagram

There is an old saying: 'A picture is worth a thousand words.'

So, if a diagram is not provided then draw one. It might be helpful, and could give you some ideas about how to solve the problem.

Draw a decent diagram. A sketch is probably fine, but it needs to look like the thing it describes. If there is a triangle in the problem, then your shape should be a triangle. If there is supposed to be a straight line then your line should be straight. The actual sizes of sides and angles are probably not important. Try to make your diagrams large and clear.

Label your diagram. If there is information provided in the question (such as the lengths of sides, or the sizes of angles) then write these on your diagram. This will often help when you are solving a problem.

Add new information that you work out. When you work out something new, add this to the diagram too.

So, in summary:

- draw a decent diagram
- label it
- add new information that you work out.

It is fairly obvious what to do when a diagram has been provided as part of a question, but a diagram can sometimes be useful in other situations. Here are two examples where drawing diagrams could help you.

> At a fast food restaurant there is a 'meal deal' that involves first choosing one of the following: cheeseburger, chicken burger, veggie burger or salad, and then ordering a side from the following list: fries, baked potato or coleslaw.
>
> How many different meals could you have?

You could work systematically and create a list, but a diagram would also help.

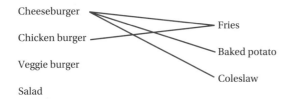

The diagram above shows all the options and the lines show some of the possible combinations.

There are three lines coming from cheeseburger. How many lines will come from chicken burger when the diagram is finished? Will this help you answer the question?

Alternatively, you could create a table like the one below to help you. What does each cell in this table represent? How does this table tell you, at a glance, how many meal possibilities there are?

	Fries	Baked potato	Coleslaw
Cheese-burger			
Chicken burger			
Veggie burger			
Salad			

Here is another example:

Jilly says, "If I write out numbers in rows of six, all of the prime numbers will either be in the top row, in the column that has 1 at the top, or in the column that has 5 at the top."

Can you tell if she is right?

1	2	3	4	5	6
7	8	9	10	11	12
13	14	15	16	17	18
19	20	21	22	23	24
25	26	27	28	29	30
31	32	33	34	35	36
37	38	39	40	41	42
43	44	45	46	47	48
49	50	51	52	53	54
55	56	57	58	59	60
61	62	63	64	65	66
67	68	69	70	71	72
73	74	75	76	77	78
79	80	81	82	83	84
85	86	87	88	89	90
91	92	93	94	95	96

You need to have some numbers to look at here so a diagram will be important.

Now you could start highlighting a few prime numbers.

This looks good so far. (It is worth noting that Jilly didn't say that every number in the top row would be prime, just that all of the primes would be in the top row or the 1 or 5 columns).

Now – why can't there be any prime numbers in the column with 2 at the top?

1	2	3	4	5	6
7	8	9	10	11	12
13	14	15	16	17	18
19	20	21	22	23	24
25	26	27	28	29	30
31	32	33	34	35	36

When you go down a row, it is the same as adding 6 to the number above. The 2 column goes: 2, 8, 14, 20, 26, and because you started with an even number, and are adding an even number, these will always be even. So, there can't be any more prime numbers in this column.

The column with 3 at the top has numbers that are all odd. But they are all multiples of 3.

The column with 4 at the top cannot have any prime numbers in it. Why not?

Neither can the column with 6 at the top.

3
9
15
21
27
33

This just leaves you with two columns and the top row for primes to go. This means that Jilly is right.

The following problems may be solved using more than one method; however, the worked solutions provided at the back of this book are based on the method introduced above.

Peter the factory manager planned to install a new hot drinks machine for the factory workers. He decided to fill it with tea because he thought tea was the most popular hot drink.

Tip

Think about what type of diagram might be helpful.

The workers did a survey to check what the preferred hot drink was among them. Each person could choose one drink from hot chocolate, tea or coffee.

Eight women wanted hot chocolate.

A total of 16 workers wanted tea, of which seven were men.

Ten men and 12 women wanted coffee. There were 25 men in total.

Was Peter correct?

Mr Rixson and Mrs Lloyd are the A-Level Mathematics teachers at Swanend Hill School. They are extremely competitive and often have debates about their students' results.

Tip

Draw a possible box plot for both classes.

The results from the latest assessments were as follows:

Class	Mean score	Median score	Range	IQR	Lowest score
Rixson	7.5	7.5	9	4	3
Lloyd	9	7	17	14.5	1

Both Mr Rixson and Mrs Lloyd have 12 students in their A-Level Mathematics classes.

Compare the results of the classes.

A rectangle has length $(2x + 3)$ and width $(x - 1)$.

a Write an expression for the perimeter of the rectangle.

b Write an expression for the area of the rectangle.

Tip

$253 = 23 \times 11$

The area of the rectangle is $250\,\text{cm}^2$.

c How long is the longest side?

d What is the perimeter of the rectangle?

In the local cement factory, the cement bags are placed on pallets made of planks of wood and bricks.

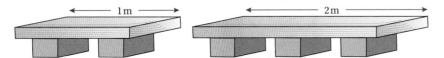

The number of bricks needed to make the pallet is calculated as 'one more than the length of the plank in metres'.

a What length of pallet uses five bricks?

b If the pallet is 7 m long, how many bricks are used in it?

The factory needs pallets with a total length of 15 m for the next batch of cement. It has planks of wood that are 4 m long and 3 m long.

c What combinations of planks can they have?

d How many bricks would be needed for each combination?

The probability Leela catches the 6:30 am train to Brighton is 0.7.

If she misses the train she will be late for work.

The probability the train will be late is 0.15.

If the train is late she will be late for work.

What is the probability Leela will be on time for work on a particular day?

Tip

What type of diagram might help?

Two five-sided spinners are numbered 1 to 5.

When the arrows are spun, your total score is calculated by adding the two numbers the spinners land on.

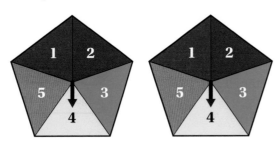

a Draw a suitable diagram to show all possible outcomes when spinning these spinners.

b What is the highest score you could get?

c What is the probability of getting a total score of 8?

Marianne needs to make a long-distance journey. She is looking for the cheapest car hire.

Whacky Wheels has a standard charge of £35, then 15p for every kilometre driven.

Wheelies Rentals has a charge of 23p per kilometre travelled.

a Complete the charges graph for both car hire companies.

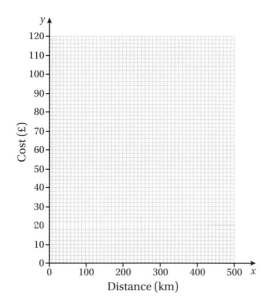

> **Tip**
>
> In this question you can use the axes that are given to help you draw the diagram.

b Marianne thinks the return journey is 300 km. Which company would be the cheaper to use?

c Marianne made a mistake in her route plan and the return journey was 500 km. How much money would Marianne have saved by using the other hire company?

The vertices of a quadrilateral are A, B, C and D.

A has coordinates (2, 1).

$\overrightarrow{AB} = \binom{2}{3}$, $\overrightarrow{BC} = \binom{4}{0}$, $\overrightarrow{AD} = \binom{4}{0}$

a Write a column vector for $\overrightarrow{CD}$.

b Compare $\overrightarrow{CD}$ with $\overrightarrow{AB}$. What do you notice? Can you say why?

c What type of quadrilateral is ABCD?

> **Tip**
>
> Draw the shape on squared paper.

Ann-Marie wants to plant a cherry tree in her garden. When it is fully grown it will have a diameter of 3 m.

Ann-Marie wants all of the fruit to fall on her lawn.

Here is a sketch, not drawn to scale, of Ann-Marie's garden.

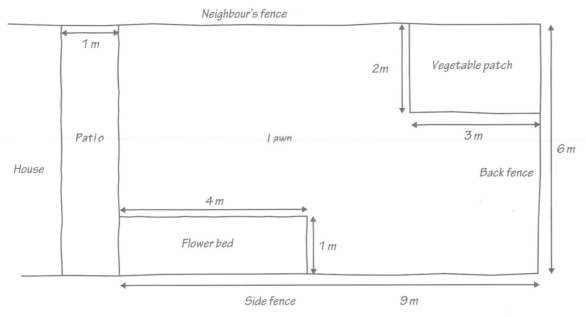

Where could the tree be planted?

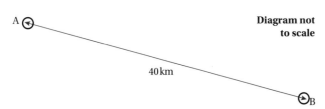

The diagram represents two remote towns A and B in the Scottish Highlands.

The mountain rescue helicopter from each town is dispatched to rescue any casualty within a radius of 25 km of that town. The fire brigade from town B will travel to any accident scene closer to town B than town A.

Draw a diagram and shade the region that the helicopter and fire brigade will both cover.

A projector is placed 1 m from a screen. When the projector is turned on, the image produced is only 20 cm high.

How far back should the projector be moved in order to produce an image that fills the screen, which is approximately 1.5 m in height?

(Assume that no other adjustments are made to the projector.)

Elspeth has an allotment. She is testing out two different types of grow bags for her tomato seedlings, which she gets to a healthy stage and then sells to her neighbours.

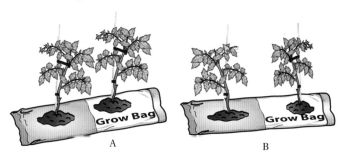

A   B

She has planted twenty seedlings in each type of bag. After three weeks she has recorded the heights of the plants as follows:

	Shortest height (cm)	Lower quartile (cm)	Median (cm)	IQR (cm)	Biggest height (cm)
Grow bag A	5	12	15	8	22
Grow bag B	8	13	15	5	25

a Compare the data.

b In your opinion, which type of grow bag is better: A or B? Give reasons for your choice.

Granny Bessie is making a patchwork quilt with scraps of fabric.

Tip

The diagram in the question is very detailed. Could a simpler diagram help?

Each patch is $(2x - 3)$ cm long and $(x + 3)$ cm wide.

a There are 25 patches in each row. Write a possible expression for the width of the quilt.

b There are 32 patches in each column. Write a possible expression for the length of the quilt.

c Write an expression for the area of the quilt, in the form $m(ax^2 + bx + c)$ where m is a constant.

The area of the quilt is $2.8 \, \text{m}^2$.

d What are the dimensions of each patch? Give your answers in centimetres.

Tip

$2.8 \, \text{m}^2 = 28\,000 \, \text{cm}^2$. It may be easier to work in centimetres.

A square-based food container has a capacity of $1440 \, \text{cm}^3$.

The base of the container is of length x cm.

a Write an equation for the height of the container in terms of x.

The inside of the container (base and four sides) is to be lined with parchment paper with no overlaps.

b What is the formula, in terms of x, for the area of parchment paper needed?

c If the height of the container is 10 cm, what is the size of the base?

d What is the area of parchment paper needed?

A rectangular swimming pool is surrounded by a path made of decorative mosaic tiles. The width of the path is x.

The pool itself measures 35 m by 30 m.

a Write an expression for the area of the tile border in terms of x.

Having the tile path laid was very expensive.

It cost a total of £3196.80, at a rate of £30 per square metre.

b Find the width of the path to the nearest centimetre.

Serrianne has taken up golf and goes to a golf range twice a week. She uses one bucket of balls each time. In every bucket of 25 balls there are always 3 yellow balls; the rest are white.

Serrianne hits one ball (chosen at random) at a time.

a What is the probability that the first 3 balls she uses will all be yellow?

b What is the probability that the first 3 balls she uses will all be white?

c Calculate the probability that the first 3 balls Serrianne uses are a mixture of two yellow and one white.

To make the journey to work Abu must drive through two sets of traffic lights.

The probability of the first set being green is 0.7. If the first set is green, the probability of the second set also being green is 0.8. But if the first set is not green, the probability of the second set being green is 0.4.

a What is the probability that Abu does not have to stop on his journey to work tomorrow?

b What is the probability that Abu only has to stop once on his journey to work tomorrow?

Geoff and Ravinder are very competitive and often have badminton and squash matches. The probability of Geoff winning at badminton is 0.85 and the probability of Geoff winning at squash is 0.35.

a What is the probability that the next time they play both matches, Geoff wins both?

b What is the probability that Geoff loses at badminton but wins at squash?

c What is the probability that both boys win one match each?

On a commercial flight to Tanzania the passengers were questioned about their malaria precautions. Only 70% of the passengers had obtained and started a course of anti-malarial tablets. The chances of getting malaria are $\frac{1}{200}$ if you take the tablets but $\frac{1}{50}$ if you are not taking the tablets. What is the probability that one passenger selected randomly will catch malaria?

Tip

What type of diagram would be helpful?

The owner of a new bookshop carried out a survey to find out the most popular A-Level courses to help decide how many revision guides to stock. A total of 200 students were asked whether they were studying Chemistry, Physics or Maths.

43 of the students surveyed did not study any of these three subjects.

A total of 92 were studying Chemistry.

There were 23 studying both Chemistry and Maths, but not Physics.

There were 19 studying both Physics and Maths, but not Chemistry.

29 were studying only Physics, and there were a total of 74 who studied Physics.

53 of the students studied two of these three subjects.

a Display the information in an appropriate diagram.

b If one person was chosen at random, what is the probability they studied only maths?

c If one person was chosen at random, what is the probability they studied at least two of the subjects?

Caroline and Janet do some swimming every morning. They swim a total of 45 lengths each. They always start together but never finish together. They swim at different speeds for different swimming strokes.

Caroline always swims 45 lengths of breaststroke in 30 minutes, completing each one at the same speed.

Janet always does 30 lengths of front crawl in the first 12 minutes, then the remaining 15 lengths at a speed of one length per 40 seconds.

a After ten minutes, how many lengths has Caroline completed?

b How long does it take Janet to complete her final 15 lengths?

c What is Caroline's speed in lengths per minute?

d How long must Janet wait for Caroline to finish?

e Roughly, on average, how many lengths does Janet swim each minute?

f If Caroline continued swimming for another 10 minutes, in theory how many lengths should she complete in total? Explain why this figure might not be correct.

The point A has coordinates (2, 2).

$\overrightarrow{AB} = \begin{pmatrix} 2 \\ 5 \end{pmatrix}$

$\overrightarrow{BC} = \begin{pmatrix} 3 \\ 0 \end{pmatrix}$

Tip

Draw a sketch of the shape.

a Find a possible pair of coordinates for D, if ABCD is an isosceles trapezium.

b Write $\overrightarrow{AC}$ as a column vector.

c Find the coordinates of E, if $\overrightarrow{AE} = 4\overrightarrow{BC}$.

d Using these coordinates for E, write $\overrightarrow{BE}$ as a column vector.

ABCD is a field surrounded by fences AB, BC, CD and DA.

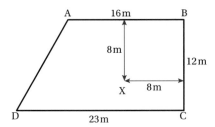

There is a dog tied to the spike X on a lead measuring 3 m.

There is a bull on an 8 m rope that is attached to the top of post A.

Find a route from corner D to corner B that would avoid both the bull and the dog.

Eve took a photo of her mum on holiday. Later, when Eve looked at the picture, she saw that her mum seemed to be the same height as a hill in the background.

Eve stood approximately 3 m away from her mum when she took the photo, and they were approximately 2 km away from the hill. Eve's mum is 15 cm taller than Eve.

Roughly how high is the hill?

When enlarging photographs, the increase in width and length must be directly proportional to each other or the photos will be distorted.

A photo has width 40 cm and length 55 cm.

a   An enlargement of this photo has width 112 cm. Find the length for this photo poster.

b   Another enlargement of the original photo has length 148.5 cm. What must the width be for this enlargement?

c   Another photo with length 15 cm and area 127.5 cm^2 is enlarged to a poster of width 25.5 cm. What is the area of this poster?

a Harriet has a challenge for her classmate Janet:

"I'm thinking of a triangle…
It has a right angle. It has one angle of 40°. It has one side that has a length of 5 cm.
Draw my triangle."

   i Accurately construct a triangle that satisfies Harriet's conditions.

   ii Demonstrate that there is more than one triangle that Harriet could be thinking of.

   iii How could Harriet alter her challenge so that only one triangle is possible?

b Janet comes up with a challenge for Harriet:

"I'm thinking of a triangle…
It has one side of length 4 cm. It has another side of length 7 cm. The angle in between these two sides is 55°.
Draw my triangle."

   i How many triangles satisfy Janet's conditions? Give reasons for your answer.

   ii Find the length of the third side.

Tip

To construct an accurate triangle you may need to use a ruler, protractor and a pair of compasses.

Tip

Before you start your accurate construction, make some sketches to show the positions of the sides and angles you are given.

You can use the three transformations listed below:

A Reflect in the line $y = x$.

B Translate by $\begin{pmatrix} 1 \\ 0 \end{pmatrix}$.

C Enlarge by scale factor $\frac{1}{2}$ about the point (2, 3).

a Carry out all three transformations, in order, on a starting shape of your choice.

b How does the resulting image change if the transformations are applied in reverse order? $C \longrightarrow B \longrightarrow A$

c How many different final images could be produced by changing the order in which the three transformations are applied?

Tip

You will find this question easier if you try it out. Think about how you can make it simpler by choosing shapes and side lengths that make the enlargement easier.

Two of the vertices of an equilateral triangle are located at points with coordinates (0, 0) and (6, 0).

a Work out the possible coordinates of the third vertex.

b If two of the vertices of a different equilateral triangle are located at (−3, 2) and (5, −4), what is its area?

29 ✪✪✪

An astronomer wants to calculate the distance to one of our closest stars, Proxima Centauri. In order to do so they take two angle measurements, six months apart. The two angles measured by the astronomer are shown in the diagrams below.

$$1\,\text{AU} \approx 1.5 \times 10^8\,\text{km}$$
$$1\,\text{arcsec} = \left(\frac{1}{3600}\right)^{\circ}$$

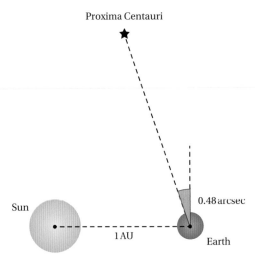

Six months later...

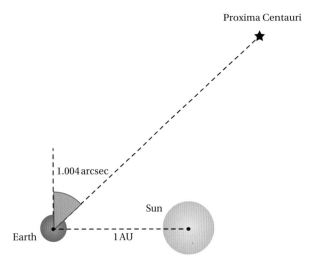

Use the astronomer's measurements to calculate the approximate distance to Proxima Centauri.

With some problems, it is easier not to focus straight away on the answer you need. Sometimes it is easier to work out want you **can** do, and this might lead you to the answer you **need** almost by accident! Here is an example to demonstrate.

In the diagram below, a square and an isosceles triangle sit on a straight line. Work out the size of angle A.

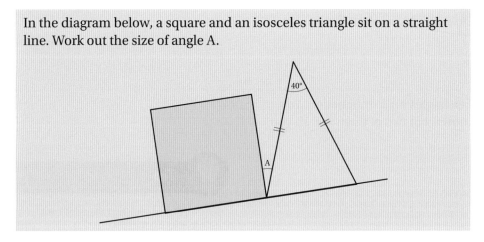

At the moment you can't work out angle A. But there are some things you **do** know.

There is a square in the diagram, so you can put in some angles.

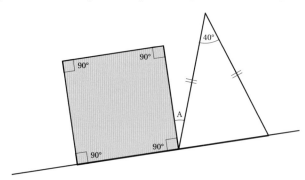

Some of these might not be helpful but, as there are only a few angles on the diagram, you are already narrowing down the problem.

Another piece of information you have got is that the triangle is isosceles. That means you can work out the other angles in the triangle. One angle is given as 40° and you know that all the angles in the triangle add up to 180°. This means the other two add up to 140° and because they are both equal they must each be 70°.

Now mark those angles on the diagram.

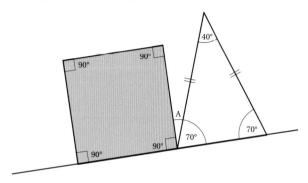

Now you can see what to do!

In the middle of the diagram there are three angles on a straight line. You know that these add up to 180° and you know what two of them are, so you can work out what angle A is.

90° + 70° = 160°

180° − 160° = 20°

So angle A = 20°

Now go back to the question.

In this diagram a square and an isosceles triangle sit on a straight line. Work out the size of angle A.

Check whether you missed out any important information. In this case, you will see that you have used all of the information provided in the question.

In this diagram a square (*you used the fact that angles in a square are 90°*) **and an isosceles triangle** (*you used the facts that base angles in an isosceles triangle are equal and that angles in a triangle add up to 180°*) **sit on a straight line** (*you used the fact that angles on a straight line add up to 180°*). **Work out the size of angle A.** (*You've done that.*)

The key points are: draw a good diagram, work out the things you can, and then write this new information on the diagram.

Here is another example.

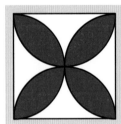 **This diagram consists of a square with four semicircles drawn inside. What fraction of the square is shaded?**

Tip

Remember: Angles on a straight line add up to 180°.

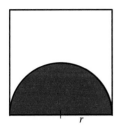

Don't worry that you can't work out the answer immediately. How about working out the area of one of the semicircles?

There aren't any measurements so you could call the radius of the circle r. The area of the semicircle is $\frac{1}{2}\pi r^2$.

If you remove a triangle as shown, you get something that is part of the original shaded diagram. The area of the triangle is r^2.

Hold on: the shaded parts of this diagram added together make one of the 'petals' in the original diagram! One petal has an area of $\frac{1}{2}\pi r^2 - r^2$.

The fraction of the original diagram (which is a square of side $2r$) that is shaded is $\dfrac{4(\frac{1}{2}\pi r^2 - r^2)}{(2r)^2}$

The numerator can be factorised to give $\dfrac{4r^2(\frac{1}{2}\pi - 1)}{4r^2}$ which is $\frac{1}{2}\pi - 1$.

You can check this is sensible by ensuring it is between 0 and 1: $\frac{1}{2}\pi - 1 \approx 0.57$

So, by starting with something that you can work out (the area of a semicricle), you have solved the problem.

The following problems may be solved using more than one method; however, the worked solutions provided at the back of this book are based on the method introduced above.

Without using a calculator, decide if the statement below is true or false.

$10\sqrt{70} < 80$

Give reasons for your answer.

The mean of five numbers is 12. The numbers are in the ratio $1:1:3:4:6$.

Find the largest number.

Jenny makes a kite. She starts from a square piece of paper, like this.

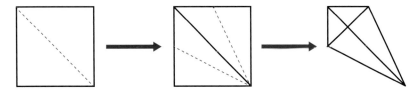

Jenny makes more kites and puts them together around a point. The start of this pattern is shown in the diagram. How many kites does she need altogether so they fit together without any gaps?

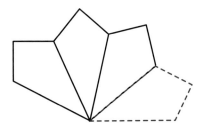

The maximum safe load of Johnny's van is 840 kg, correct to the nearest 10 kg.

Johnny is going to transport large stone slabs for his next garden project.

The slabs weigh 16 kg each correct to the nearest kilogram.

Johnny doesn't want to risk damaging his van.

What is the maximum number of slabs Johnny can safely transport in one go in his van?

a Find a formula in terms of x for the area of the shape shown below.

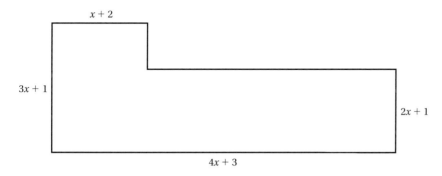

The actual length of the side labelled $3x + 1$ is 16 m.

b What is the area of the shape? Give your answer in square metres.

The shape is the plan of a field. Farmer Smith wants to put a fence around the field.

He intends to put three lengths of wire between wooden posts around the field, as shown in the diagram.

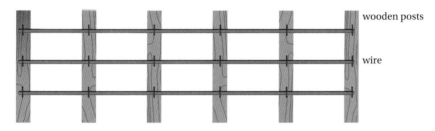

Farmer Smith will fix the wooden posts one metre apart.

c How many wooden posts will Farmer Smith need in total?

d What length of wire will Farmer Smith require for three lengths, as shown in the diagram?

Posts cost £18.50 each and wire is sold at £2.30 per metre.

e How much is it going to cost Farmer Smith to fence the field?

6

Benevolence paid £2.90 for three mangoes and two pawpaw fruits at the market.

Cecilia paid £5.50 for five mangoes and four pawpaw fruits at the same market.

What is the price of a mango and the price of a pawpaw fruit?

7

A bag contains green, yellow and white balls. The probability of picking a green ball out of the bag is 0.64. The probability of picking a white ball is the same as the probability of picking a yellow ball.

a What is the probability of picking a white ball?

b If there are nine yellow balls, what is the total number of balls in the bag?

8

Sharon, Sue, Frances and Anisha are all running in a 5 km sponsored race in Cambridge.

The probability of each of them finishing first out of the four of them is shown in the table below:

Sharon	0.23
Sue	0.46
Frances	0.15
Anisha	x

What is the probability that either Sue or Anisha finishes first?

The teacher has a dice but doesn't know whether it is biased.
20 children each throw the dice 10 times. The number of times each
child rolls a six is shown below.

Is the dice biased? Give reasons for your answer.

Child	Number of sixes in 10 rolls
A	7
B	4
C	3
D	6
E	4
F	3
G	4
H	5
I	5
J	3
K	4
L	5
M	6
N	4
O	5
P	7
Q	0
R	3
S	3
T	2

Sandra wants to make a triangular prism:

She draws the diagram below, cuts it out, folds it up and tapes it together.

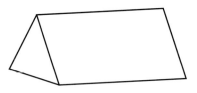

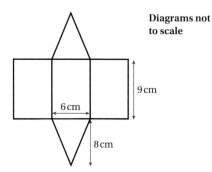

Diagrams not to scale

9 cm

6 cm

8 cm

a What is the volume of the triangular prism?

Chris makes a different prism:

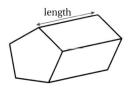

length

The pentagonal faces each have an area of 36 cm².

The volume of Chris's prism is the same as the volume of Sandra's prism.

b Calculate the length of Chris's prism.

Jeff takes a full 500 ml bottle of water and pours himself a drink into the glass shown below. He fills the glass to 2 cm below the top.
How much water is left in the bottle?

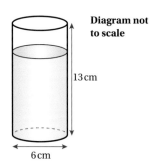

Diagram not to scale

13 cm

6 cm

12

Mrs Jones and Miss Smith are writing topic tests for the new scheme of work for GCSE Mathematics.

Each topic test is going to have 32 questions split in the ratio of 3:5 for non-calculator to calculator questions.

a How many non-calculator questions will each topic test contain?

There will eventually be 15 topic tests.

b How many calculator questions in total must be written?

Miss Smith will write all of the non-calculator questions for the 15 tests and some of the calculator questions so that she and Mrs Jones will each write the same number of questions.

c What fraction of the total questions written by Miss Smith will be calculator questions?

13

$\frac{4}{5}$ of the rectangle on the right is shaded.

a How many extra unit squares must be shaded in order to increase this fraction to $\frac{5}{6}$?

The $\frac{4}{5}$-shaded rectangle is put next to a congruent rectangle that is $\frac{2}{3}$ shaded.

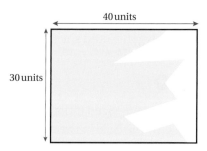

b What fraction of the resulting large rectangle is shaded?

c How many square units need to be shaded to increase the fraction so that exactly $\frac{3}{4}$ of the large rectangle is shaded?

Tip

You could draw a diagram showing the shading in a 'neater' way.

14

Your younger brother uses his calculator to work out $6 \div 0.5$ and is surprised that he gets such a big answer. He tries $6 \div 0$ and doesn't understand why it won't tell him the answer.

How could you talk about the answers to your brother?

Tip

Think about how you might do this in a way that someone younger than you would understand. Remember, you have to be clear enough to change their mind!

Alison is planning to make a patchwork quilt for her niece. She has decided to include the flower pattern shown below as part of her design. The centre piece of this flower is a regular hexagon.

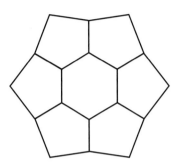

a Show that Alison cannot produce this pattern by using regular pentagons.

b She still wants to use a regular hexagon for the centre piece. Are there any regular polygons that she could put around the edge to form the petals?

c If she wanted to use regular pentagons for the petals, which regular polygon would she need in the centre?

Tip

What do you know about a regular hexagon? What angles can you work out? What about a regular pentagon?

An average-sized toilet paper roll is 10.5 cm wide and has a diameter of 12 cm. The hole in the centre has a diameter of 5 cm.

A good-quality paper has thickness 0.5 mm and each sheet is 10 cm long.

'Happy Wipes' advertise their toilet rolls as the best on the market, with each roll containing at least 200 good-quality sheets.

Is their claim true?

Tip

What can you work out? Can you find the volume of a sheet of toilet paper? What about the volume of the toilet roll?

Jo creates a quadrilateral by folding a rectangular piece of paper as shown in the diagram below.

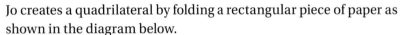
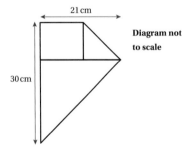

Diagram not to scale

21 cm

30 cm

a What is the area of the quadrilateral?

b What fraction of the original rectangle does the quadrilateral represent?

Tip

For part **b** you could give your answer as a fraction, decimal or percentage. If you choose to use a fraction, make sure you give it in its simplest form.

The area of the rectangle is the same as the area of the square.
The dimensions of both shapes are in centimetres.

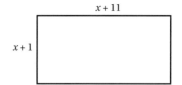 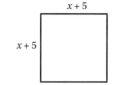

Diagrams not to scale

How many centimetre squares are needed to cover the rectangle?

A shape is formed by cutting a square from the corner of a larger square, as shown in the diagram.

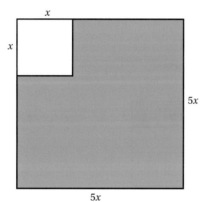

Diagram not to scale

a Write an expression for the blue area.

b If the area of the blue shape is 1944 cm², find the perimeter of the blue shape.

In the diagram, AC is a diameter of the circle and AD = BC.

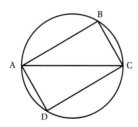

 Tip

What do you know about the triangles? Consider how knowing about congruent triangles might help you with this question.

Show that quadrilateral ABCD is a rectangle.

A trapezium is inscribed in a circle with centre O.

∠BOC is 50° and ∠AOB is 75°.

a Calculate the size of ∠COD.

b What does this tell you about trapezium ABCD?

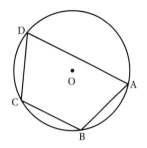

Diagram not to scale

Harriet says: "I think tan θ is equal to sin θ ÷ cos θ."

Using the values in the table below to give reasons for your response, decide whether Harriet is correct.

θ	$\sin \theta$	$\cos \theta$	$\tan \theta$
0°	0	1	
30°	$\frac{1}{2}$	$\frac{\sqrt{3}}{2}$	
45°	$\frac{\sqrt{2}}{2}$	$\frac{\sqrt{2}}{2}$	
60°	$\frac{\sqrt{3}}{2}$	$\frac{1}{2}$	
90°	1	0	

Sunil has kept woodlice as his pets for a year. They have survived well in their plastic habitat, but the woodlice expert at the pet shop told him he should only keep approximately 150 woodlice in a habitat of that size. Sunil thinks it is getting close to the time when he should let some go free, so he decides to count them by using a capture–mark–release counting method.

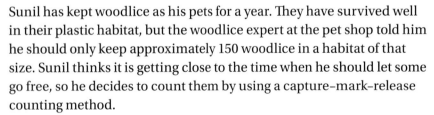

He catches 16 woodlice and marks each one with a dot. He then places them back in their habitat. Two days later Sunil catches 16 of his woodlice again and discovers he has recaptured two with the dots.

On the basis of this information, does he need to set any free?

Five equilateral triangles are drawn on the edges of a regular pentagon.

Show that the polygon created by joining points F, G, H, I and J is also a regular pentagon.

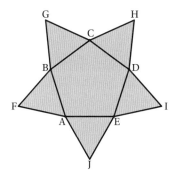

Many astronomical distances are measured in light years. A light year is the distance that light travels in a vacuum in one year.

Light travels at approximately 2.99×10^5 km/s.

The distance to Proxima Centauri is approximately 4.22 light years.

How many laps of a standard 400 m running track is this equivalent to?

Tip

What can you work out? How far does light travel in a minute/hour/year?

Mary wants to make a medieval-style hat for her Halloween ball. She will buy a sheet of black card, big enough for her to be able to make the hat without joining any pieces of card together.

The hat must be made from one piece of card.

The hat will be conical in shape with a radius of 9 cm and slant height of 36 cm.

What size of card must Mary buy?

Tip

The net of this hat is a sector.

Tomato soup contains 59 kilocalories per 100 ml. Martin is on a diet and has been advised to use a smaller bowl to help reduce his portion sizes. The images show Martin's original bowl and the smaller bowl he will now use. Martin always fills his bowl as close to the brim as possible.

Assuming the two bowls are perfect hemispheres, how many kilocalories will he save by eating tomato soup from the smaller bowl? Give your answer to the nearest kilocalorie.

Look at the two triangles drawn on this coordinate grid.

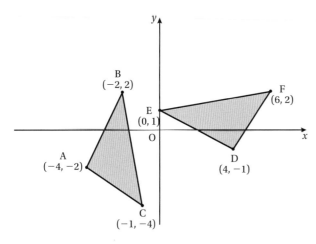

Show that the two triangles are congruent.

> **Tip**
>
> You can't say straight away if they are congruent, but you can calculate the lengths of some sides.
> Remember to refer to the conditions for congruence.

The hare and the tortoise are having a race. The hare is confident and offers the tortoise a shortcut across the middle of the rectangular field.

If the tortoise totters along at 0.1 m/s, what is the lowest speed the hare could run and still win the race?

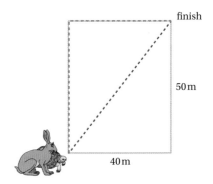

3 Adding lines

Adding lines is a strategy that works for some geometry problems. Sometimes an extra line is very helpful. On other occasions it turns out to be useless. How can you tell which type of line to use? Let's work through an example problem.

In the diagram below, two parallel lines are shown. Work out the size of angle *a*.

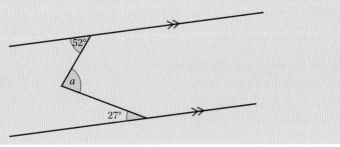

There are several different ways to solve this problem. Here are three of them, all of which involve adding extra lines to the diagram.

Solution 1

Add a line that is parallel to the other two lines and that passes through the angle you want to work out:

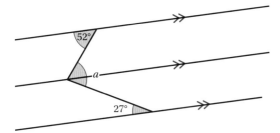

Could this help you?

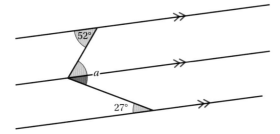

The angle shown in red is equal to 27°. Can you say why?

What is the other part of angle *a* equal to? What is angle *a*?

Solution 2

Add a line that is perpendicular to the parallel lines and that passes through the point of the angle, as shown in the next diagram.

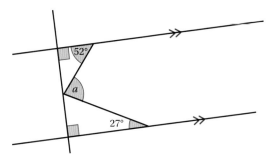

You now have two triangles. In those triangles you know two of the angles, so you can work out the third one. Do that now, remembering to mark the angles on the diagram. Here you are making use of the strategy from Chapter 2 and advice from Chapter 1 to annotate a diagram by adding new information to it when you work it out.

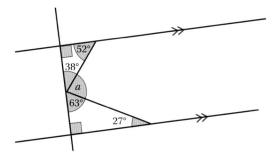

In the middle of this diagram there is a straight line, so now you can work out the missing angle a.

Solution 3

Add a line that is perpendicular to the parallel lines and that passes through the point of the 27° angle as shown in the diagram here.

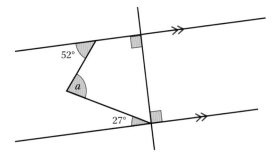

You can see a quadrilateral in the middle of the diagram. You can also 'see' what to do to solve the problem, because you can work out three of the angles in the quadrilateral and you know that all of the angles add up to 360°.

If we go back to the question we posed at the beginning: How can you tell which type of line to use?

The short answer is that you can't. There are lots of different lines you could have drawn on this diagram that wouldn't have helped you. For example, this line in the next diagram doesn't do anything special and is not helpful.

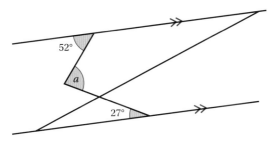

A line that is 'special' in some way is more likely to be helpful.

Try using lines that:

- are parallel to other lines
- are perpendicular to other lines
- are lines of symmetry
- join up two points
- are midpoints of sides of shapes.

Tip

You may need to try out several ideas before one works!

The following problems may be solved using more than one method; however, the worked solutions provided at the back of this book are based on the method introduced above.

Students taking part in a mathematical challenge are shown this image.

They are asked the question 'What proportion of the outer (large) triangle is coloured purple?'

After some thought, the teams hand in their answers. Four of these answers are shown below:

Team A: 24 cm²

Team B: $\frac{3}{4}$

Team C: $\left(\frac{3}{4}\right)^2$

Team D: 0.5625

The question is marked out of 4 marks.

How many marks would you award each team?

Give reasons for your decisions.

2

Robyn is designing a template for an art project. She plans to cut a regular octagon in half through two vertices and place them side by side, as shown.

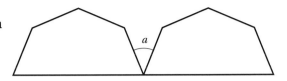

Robyn is trying to calculate the size of the angle labelled a.

Show two ways in which Robyn could do this.

3

Mandeep draws eight dots equally spaced around a circle.

He draws a triangle by connecting three of the dots. The line AC passes through the centre of the circle, O.

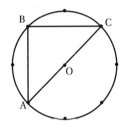

Diagrams not to scale

> **Tip**
>
> Drawing extra lines on the diagrams can help you to answer questions like this. Think carefully which line to draw.

a Use this image to show that ∠ABC is a right angle.

Mandeep draws another triangle by connecting a different set of three dots. The line AC still passes through the centre of the circle.

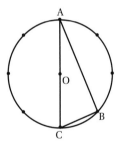

b Is ∠ABC a right angle?

Mandeep draws a different circle and marks ten dots equally spaced around its circumference. Once again, he draws a triangle by connecting three of these dots. The line AC passes through the centre of the circle.

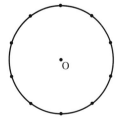

c Show that, for a triangle of your choice that meets these conditions, the sum of ∠BAC and ∠BCA is 90°.

Liz looks up at the clock on her office wall. She is disappointed to see that it is not yet five o'clock, which is the time she can go home.

a Through how many degrees must the clock's minute hand turn before Liz can leave work?

b Through what angle must the hour hand turn before Liz can leave work?

Tip

Remember that, on an analogue clock, the hour hand and the minute hand both move continuously. What angle does each hand turn through in one hour?

Gerry needs to put a short fence around his swimming pool to prevent small children falling in.

He wants the fence to surround the pool completely, at a distance of 1 m from the edge.

a Calculate the length of fencing that Gerry will need to buy. Give your answer correct to 3 sf (significant figures).

Ian also needs to build a fence around his pool, but it is a different shape from Gerry's. Ian would like to build his fence 1.5 m from the edge of the pool.

b Will Ian need to buy more fencing than Gerry? Give reasons for your answer.

Gerry's pool

15 m

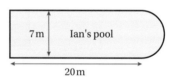

7 m   Ian's pool

20 m

The perimeter of the shape shown here is 2.28 km.

Calculate the value of *a*. Give your answer in kilometres (km).

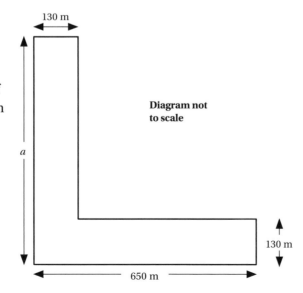

130 m

Diagram not to scale

a

130 m

650 m

Tip

Remember: 1 km = 1000 m.

A circle is inscribed in a triangle as shown below. What is the value of *x*?

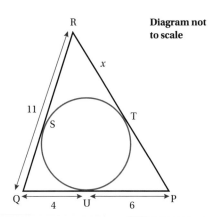

R

Diagram not to scale

x

11

S

T

Q   4   U   6   P

Tip

Join the centre of the circle to the three points of contact with the tangents.

Show how this shape could be divided into:

a three congruent shapes

b four congruent shapes.

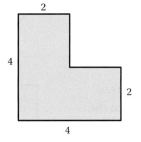

2

4

2

4

Tip

Make several copies of the diagram and try out your ideas.

Charlie's house has two floors connected by a staircase. The vertical distance between the two floors is 3.6 m. Each step in the staircase is 18 cm high and 28 cm deep.

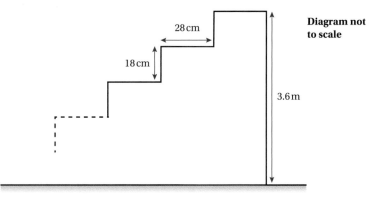

28 cm

18 cm

Diagram not to scale

3.6 m

a How many steps are there in the staircase?

b How long will the bannister rail be?

A rectangular tabletop has a perimeter of 12 m. The width of the table is x m.

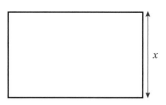

a Write an expression in terms of x for the surface area of the tabletop.

The area of the tabletop is 6.3 m².

b Would a tablecloth measuring 2.2 m by 3.1 m cover this tabletop?

Lorna is going to Glasgow for a fortnight of Scottish music festivals.

Lorna is travelling from London to Glasgow. There are a couple of different options for her journey.

Lorna can go by Gray Hound coaches, which travel direct from London to Glasgow. Alternatively, she can go by train to Glasgow with a change in Birmingham and a wait of one hour 45 minutes.

The distance from London Victoria to Glasgow Buchanan Station is 650 km.

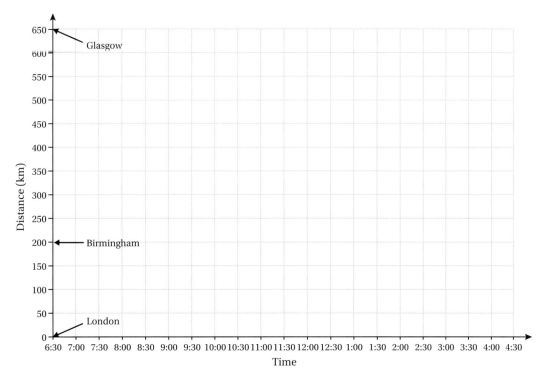

The Gray Hound coaches leave at 6:30 am and take exactly eight hours to arrive in Glasgow.

If Lorna goes by train, she would leave Victoria at 7 am and arrive in Birmingham at 8:40 am.

The train journey from Birmingham to Glasgow should then take four hours and 15 minutes.

a What is the average speed for the coach journey?

b What is the average speed for the train journey?

As shown in the diagram,

$\overrightarrow{SW} = 3\mathbf{v}$, $\overrightarrow{ST} = 2\mathbf{u}$

$\overrightarrow{SR} = -\frac{1}{2}\mathbf{u}$, $\overrightarrow{SX} = -2\mathbf{v}$

$\overrightarrow{SP}$ is the diagonal of parallelogram SWPT and $\overrightarrow{SZ}$ is the diagonal of parallelogram SRZX.

Are Z, S and P collinear?

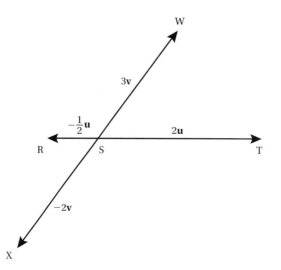

> **Tip**
>
> Drawing a clear diagram will help you see what is happening here.

A giant paperclip is placed against a ruler as shown below. The curved ends are semicircles.

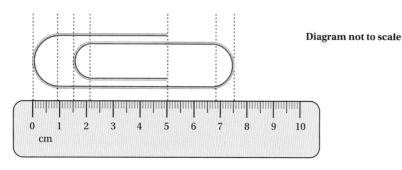

Diagram not to scale

a Calculate the length of wire needed to make this paperclip. Give your answer correct to 1 dp (decimal point).

A second paperclip is placed against the ruler.

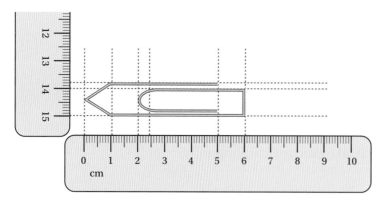

b Which paperclip uses the greater length of wire?

> **Tip**
>
> You may need to use Pythagoras' theorem in this question.

In an isosceles triangle the lengths of the sides are given as $(2x - 1)$ cm and $(2x - 4)$ cm.

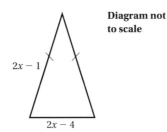

Diagram not to scale

$2x - 1$

$2x - 4$

Tip

Which topic will help you to work out the height?

a Work out an algebraic expression for the perpendicular height of this triangle.

b If the perpendicular height of the triangle is 12 cm, find the area of the triangle.

Students on a field trip are told that they can use a clever trick to estimate the height of a tall tree. They are told to point their arm at an angle of approximately 45° and to look along it. They then walk backwards until their arm is pointing to the top of the tree.

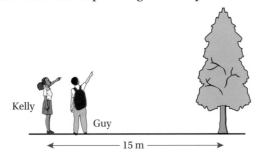

Kelly

Guy

15 m

The distance from the student to the tree trunk is approximately the same as the height of the tree, plus the student's own height.

a Say why this method gives a rough answer.

Two students, Kelly and Guy (who are the same height), try this out. Guy ends up standing 3 m in front of Kelly.

b If Kelly's angle is exactly 45°, at what angle must Guy be holding his arm?

The graphs of two quadratic equations are shown below.

The blue parabola is symmetrical about the line $x = 2$ and has roots at A and B.

The blue parabola is reflected in the x-axis to create the red parabola. The red parabola has roots at C and D.

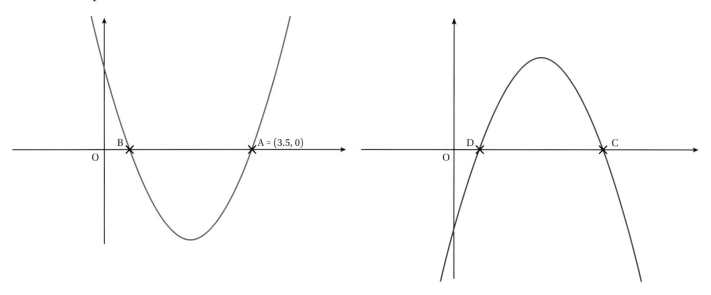

Work out the coordinates of points B, C and D.

Six cylindrical candles are held together in a plastic sleeve in the formation shown. Each candle has a radius of 2.7 cm.

Calculate the length of plastic needed to hold the candles tightly. Give your answer correct to 2 dp.

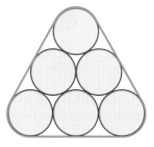

 Tip

Think about the plastic being wrapped around just one candle. What about two candles?

Two straight lines are drawn from a point A outside a circle to two points, B and C, on the circumference of the circle. The lengths of the four arcs created are marked in the diagram in terms of x.

Calculate the size of $\angle$BAC.

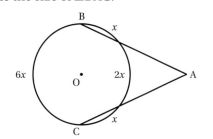

Diagram not to scale

 Tip

Are there any useful lines that you can draw to make this problem look more 'friendly'?
Why have you been told the arc lengths? What does this have to do with angles in the circle?

ABCDEF is a regular hexagon.

A has coordinates (2, 1).

The column vector for $\overrightarrow{AB} = \begin{pmatrix} 0 \\ 2 \end{pmatrix}$

a What is the column vector for $\overrightarrow{DE}$?

b Find the coordinates of C.

c What is the column vector for $\overrightarrow{BC}$?

d What is the column vector for $\overrightarrow{FC}$?

e What is the column vector for $\overrightarrow{AE}$?

f What is the column vector for $\overrightarrow{AC}$?

g What is the column vector for $\overrightarrow{EF}$?

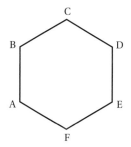

Tip

Some of your answers will involve surds.

 Two triangles are shown below.

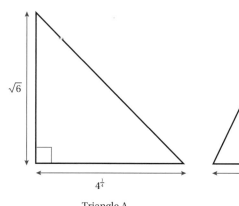

Triangle A

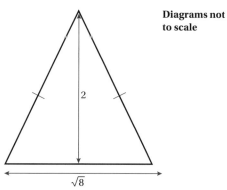

Diagrams not to scale

Triangle B

Without using a calculator, say which triangle has:

a the greater perimeter   b the greater area.

A square is inscribed inside a circle of radius 1 cm as shown.

a Calculate the area of the square.

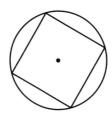

An equilateral triangle is inscribed in a circle.

b Show that the ratio of the area of the triangle to the area of the square in part **a** is approximately 2:3.

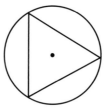

A regular pentagon is inscribed in a circle.

c Show that the area of the pentagon is approximately $\frac{6}{5}$ the area of the square in part **a**.

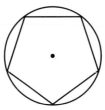

The circle has radius 5 and is centred on the origin. The point (3, 4) lies on the circle.

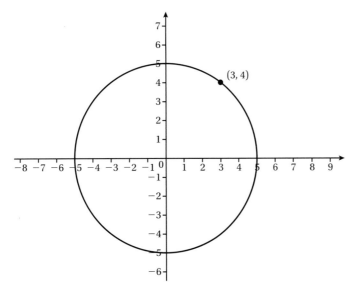

a Find the equation of the tangent to the circle at this point.

b Use your answer to part **a** to write down the equation of another tangent to the circle.

Sometimes there is a problem that is just too complicated to solve immediately. One way to deal with this is to make the problem simpler. This won't directly tell you the answer because it will be a slightly different problem, but it might give you a starting point that will help you get to the answer. Here's an example.

> George climbs a flight of 10 stairs. He can go from one stair to the next one (1-step), or can miss out a stair and go up two at once (2-step). How many different ways can he go up the 10 stairs?

You could start by drawing a diagram, but you will quickly produce a picture that has too many lines on it.

You could also start to write down the different possibilities, using 1 to stand for '1-step' and 2 to stand for '2-step'.

So:

1111111111

1211212

22222

and so on.

But there are too many possibilities and it just looks too hard.

You could make the problem simpler. Let's make it very simple indeed. Imagine that George wants to climb a flight of stairs that has 1 stair in it. This is easy: there is 1 way (if he tries to do a 2-step he will only actually go up one step). Now try a flight with 2 stairs in it. There are two ways. The diagram shows the working.

You could now look at the pattern of answers to see if that will help, as shown in this table.

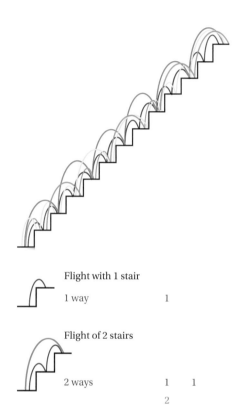

Flight with 1 stair

1 way   1

Flight of 2 stairs

2 ways   1   1

                         2

Number of stairs	Number of ways of climbing them
1	1
2	2
3	3
4	5
5	8

Do you recognise the number sequence in the second column?

1, 2, 3, 5, 8 is part of the Fibonacci sequence, where you add the two previous numbers to get the next one. 1 + 2 = 3, 2 + 3 = 5, 3 + 5 = 8. If this is the right sequence then the next one will be 5 + 8 = 13. You don't want to draw this out, so instead think about why this might be sensible.

When you look at the numbers written next to each flight of stairs you can see some similarities.

Here are the routes up the 5-stair flight:	**If you remove the final number you get this sequence:**
11111	1111
2111	211
221	22
212	21
1211	121
122	12
1112	111
1121	112

←————**and these are the ways of climbing a 3-stair and a 4-stair flight!**

Why does that make sense?

Well, if you want to climb a 5-stair flight of stairs, you could go up 4 stairs and then do a 1-step to get to the top. So you need all of the ways of climbing a 4-stair flight and can then put a 1 at the end. Alternatively, you can get to stair 3 and then do a 2-step to get to the top, so you need all of the ways of climbing a 3-stair flight with a 2 after them. This means you really are adding the previous two numbers together.

To climb a 10-stair flight you need to continue the sequence and get: 1, 2, 3, 5, 8, 13, 21, 34, 55, 89. So, there are 89 ways to climb the ten stairs.

Here, simplifying the problem meant you could get started and could see what was going on.

Let's look at another example problem.

> What is the smallest number that can be divided (with no remainder) by all of the numbers from 1 to 20?

What do you know already?

Clearly the number will have to end with a zero (to make it divisible by 10). It will have to be even so it is divisible by 2 (but if it ends with a zero then that is already the case).

One strategy that might be useful is to make the problem simpler.

Change the problem so it now says this:

What is the smallest number that can be divided by all of the numbers from 1 to 3?

If you do $2 \times 3 = 6$ then you know that can be divided by 1, by 2 and by 3. Is it the smallest number? Yes.

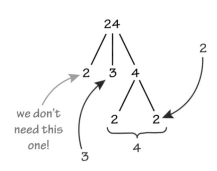

we don't need this one!

Another change:

What is the smallest number that can be divided by all of the numbers from 1 to 4?

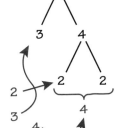

If you do $2 \times 3 \times 4 = 24$ then you know that can be divided by 1, by 2, by 3 and by 4. But there is a smaller number that works. 12 can be divided by 1, by 2, by 3 and by 4. Why didn't $2 \times 3 \times 4$ give you the right answer? To make it divisible by 4 you didn't need to multiply the previous answer by 4, because there was already a 2 involved. If you had just multiplied that previous answer by 2 that would have given you something that is divisible by 4.

Therefore the answer is $2 \times 3 \times 2 = 12$

Do you notice that these numbers are prime numbers?

Try the next one:

What is the smallest number that can be divided by all of the numbers from 1 to 5?

You currently have $2 \times 3 \times 2$, which is not helpful if you want it to be divisible by 5. The trick that you used with 4 (splitting it into 2×2) won't work here because 5 is a prime number. To make it a multiple of 5 you will need to multiply by 5, giving you $2 \times 3 \times 2 \times 5 = 60$

You can see that the next one is easy.

What is the smallest number that can be divided by all of the numbers from 1 to 6?

60 is obviously already divisible by 6, and you can see that in the numbers because 2×3 is in there. $2 \times 3 \times 2 \times 5$ is divisible by all of the numbers from 1 to 6.

To make the smallest number divisible by 7 you need to include a 7, because you cannot make 7 by multiplying smaller numbers (7 is prime).

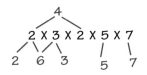

To make it divisible by 7 you therefore need to have $2 \times 3 \times 2 \times 5 \times 7$

To make it divisible by 8 you only need an extra 2 because it is already divisible by 4, and to make it divisible by 9 you need an extra 3.

Then this is already divisible by 10.

Now you can finish off the problem by thinking about each of the numbers from 11 up to 20 as shown in the table below.

11	Needed, prime number	**16**	You have 8, so just need extra 2
12	Have already ($2 \times 2 \times 3$)	**17**	Needed, prime number
13	Needed, prime number	**18**	Have already ($2 \times 3 \times 3$)
14	Have already (2×7)	**19**	Needed, prime number
15	Have already (3×5)	**20**	Have already ($2 \times 2 \times 5$)

Your final answer is therefore:

$2 \times 2 \times 2 \times 2 \times 3 \times 3 \times 5 \times 7 \times 11 \times 13 \times 17 \times 19 = 232\,792\,560$

The following problems may be solved using more than one method; however, the worked solutions provided at the back of this book are based on the method introduced above.

Anna knows that she can create a heart shape from a triangle and two semicircles. She makes a Valentine's greetings card by cutting a heart shape out of a rectangle of red cardboard 9 cm by 13 cm, as shown below.

Tip

Start simply. Find the radius of the circle first. Then can you find the areas you need?

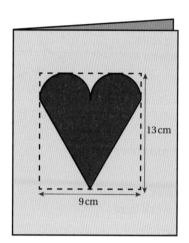

What area of red cardboard has Anna removed from the original piece to make her card? Give your answer correct to 3 sf.

Ahmed buys a piece of wood measuring 12 cm by 5 cm by 1 m for £5.

He uses the wood to make door wedges like this.

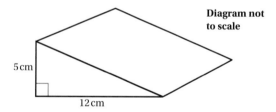

Diagram not to scale

5 cm

12 cm

a He made 40 wedges from the piece of wood he bought.
What is the width, x cm, of these wedges?

b What is the lowest price Ahmed could sell the wedges for to make at least £30 profit?

Larry's aunt gives him £40 000 to put into a compound interest savings account.

ABC Bank is offering 8 per cent compound interest on your money if you invest more than £30 000 and it remains in the bank for at least ten years.

Larry's aunt tells him that his money will more than double if he invests his money for ten years.

a Is Larry's aunt correct?

b What is the exact amount Larry will have in his account at the end of ten years?

Larry's aunt decides to invest some of her own money for ten years. She wants to make a profit and have a total of at least £75 000 at the end of the ten-year period.

c What is the smallest amount (to the nearest £1000) she must invest to have at least £75 000 at the end of ten years?

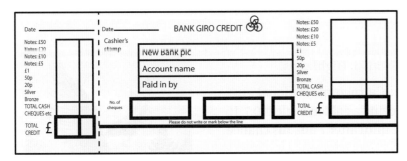

Is it possible for you to count from one to one million in your lifetime?

Tip

Don't try this out! What easier version of this question could you try out? There isn't an exact counting speed, so you won't get an exact answer!

Consider each of the calculations below. Without using a calculator decide whether the answer is odd or even. Give reasons for your answer.

a $4^3 + 3^4$ **b** $6^7 + 3^7$

Helen drives to work every day. On her journey there is a level crossing, a set of traffic lights and a roundabout. At each of these either she must stop or she is allowed to continue.

The probability of stopping at the level crossing is $\frac{1}{10}$.

The probability of stopping at the lights is $\frac{1}{4}$.

The probability of stopping at the roundabout is $\frac{2}{3}$.

What is the probability that on Helen's next journey to work she has to stop at least once?

Hamish loves to go eagle-spotting in the Scottish Highlands.

If he wakes up early and is able to be in the Cairngorms National Park before 7 am, the probability of him seeing a golden eagle is 0.7. If Hamish wakes up late and does not arrive until after 7 am, his chance of spotting a golden eagle is 0.4.

The probability of Hamish waking up early on the morning of one of his planned trips is 0.85.

a What is the probability that Hamish will see a golden eagle on one of his trips?

Sometimes Hamish takes his younger sister, Elspeth, with him on his trip to the Highlands.

The probability of Elspeth waking up early enough to allow them to arrive before 7 am is 0.75.

b What is the probability that Hamish will **not** see a golden eagle when his sister is on the trip with him?

c What happens to Hamish's chances of seeing a golden eagle if Elspeth goes with him?

Dave is travelling 280 km from Milton Keynes to Weymouth. He is travelling by car, and a graph of his journey is shown below.

Hazel is travelling from Weymouth to Milton Keynes on the same day as Dave, but she rides a motorbike.

A graph of Hazel's journey is also shown.

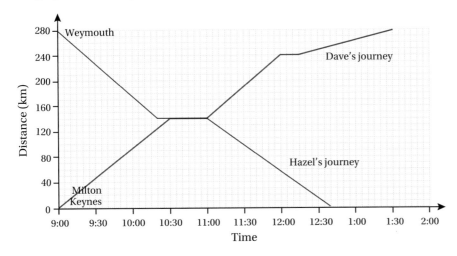

Both Dave and Hazel set out on their journeys at 9:00 am

a What was Dave's speed for the first 140 km?

b What was Hazel's speed for the first stage of her journey?

c How long did Hazel have to wait before Dave arrived at the service station where they were meeting for a coffee?

d Dave got caught up in a queue following an accident with a tractor just outside of Weymouth. What time was this?

e How long did Dave's journey take? What was his average speed?

f How long did Hazel's journey take? What was her average speed?

In the UK the national speed limit on motorways is 70 miles per hour. To convert kilometres to miles, divide by 8 and multiply by 5.

g Should either Dave or Hazel receive a speeding ticket for any stage of their journeys?

Two almost identical logos are shown below. Both show a square of side length 10 cm with a second square tilted inside it.

a Do they have the same total line length?

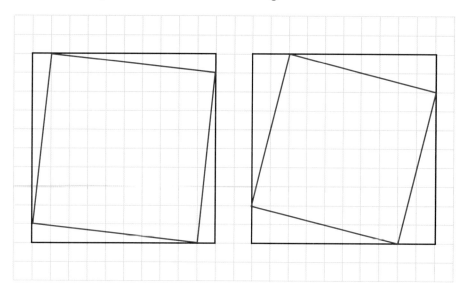

A third logo is created in the same style. The tilted square in this logo has an area exactly half that of the original square.

b Make an accurate drawing of this logo on squared paper.

Who is the oldest?

Shirley — 14.2 years

Ross — 169 months

Jemima — 5293 days

Dave — 450 million seconds

Greg is starting a bakery. He wants to advertise nutritional information on his homemade cakes.

He follows this recipe and increases it to make a batch of 30 muffins:

<u>Makes 12</u>

2 eggs   200 g caster sugar   250 ml milk

125 ml vegetable oil   400 g flour   1 tsp salt

1 Egg
Fat: 4.6 g
Sugar: Trace

Milk
Fat: 2%
Sugar: 5%

Vegetable oil
Fat: 100%
Sugar: 0%

Flour
Fat: $\dfrac{7}{500}$
Sugar: $\dfrac{1}{1000}$

Greg knows that 1 ml of liquid weighs about 1 g.

a What is the amount, in grams, of sugar and fat in each muffin?

The GDA (guideline daily amount) of fat is 70 g and sugar is 90 g.

b What should Greg advertise as the percentage of the GDA in each of his muffins?

The area of rectangle A is the same as the area of rectangle B.

Find the value of *a*.

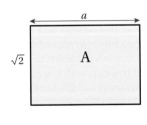

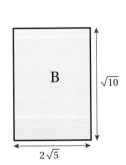

Diagrams not to scale

✪✪✪

A small rectangle is cut out of a larger rectangle and thrown away. Calculate the proportion of the original rectangle that is left.

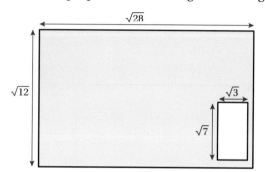

√28

Diagram not to scale

√12

√3

√7

 Tip

How would you solve it if the numbers weren't surds?

  ✪✪✪

A garden centre sells two types of hanging basket.

Tip

The right-hand basket is a hemisphere. What is a hemisphere? How would you calculate its volume?

Assuming that both types have the same diameter and that the height of the conical basket is also equal to its diameter, which basket has the greater volume?

  ✪✪✪

A mug is printed with a chequerboard pattern made up of squares that have been rotated through 45°. Exactly five of these white squares fit around the mug. The diameter of the mug is 9 cm. Calculate the perimeter of one square on the mug.

Give your answer correct to 3 sf.

 Tip

Think of the design as a rectangle that has been wrapped around the mug. Draw a diagram to show the net of the cup with the squares on it. How long will the rectangle be?

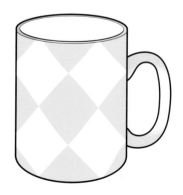

Making changes to the problem is another strategy that works particularly well with some geometric problems. This time you won't necessarily simplify the problem (like you did in the previous chapter) but you could instead tweak it in a way that doesn't change the important features of the problem. This might allow you to see what is going on.

> In this diagram AB is the diameter of the circle, C lies on the circumference and O is the centre of the circle. What is the relationship between the area of triangle OAC and the area of triangle OBC? Give reasons for your answer.

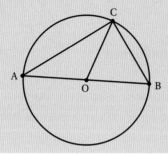

The question asks 'What is the relationship?' What could that mean? Perhaps one is double the other? Maybe they are the same?

In this diagram it looks as if the areas might be the same. Can you provide evidence to convince yourself that this is the case? Can you convince someone else?

The key points are:

- there is a circle
- AB is the diameter of the circle
- O is the centre of the circle
- points A, B and C are on the circle.

All of this must stay true. You might be able to move some of the points so the answer is obvious. There doesn't seem to be anything special about the placing of point C, so it might be acceptable to move the points around. It wouldn't be acceptable to move C off the circle, though, because that would change the scenario.

This diagram looks good. C has moved so that it is halfway between A and B but still on the circle. There is now a line of symmetry in the diagram and the two triangles are clearly equal.

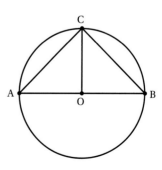

Here is another special case.

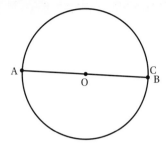

C has moved so that it is on top of point B. This time the two triangles both have zero area, so they are the same.

The idea that the two areas are the same is looking plausible.

How do you work out the area of a triangle?

It is half of the base multiplied by the perpendicular height.

In the right-hand triangle, the area is the radius of the circle (OB) multiplied by the dotted line (the perpendicular height), divided by 2.

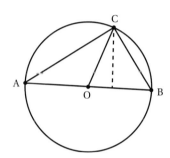

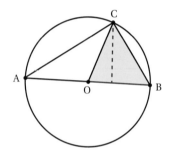

In the left-hand triangle, the area is the radius of the circle (AO) multiplied by the perpendicular height, divided by 2. But what is the perpendicular height for the left-hand triangle? It is actually the dotted line.

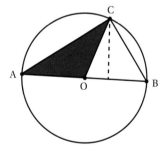

So they do have the same area and you can say why.

A good diagram is important here: see Chapter 1. This is also an example of where adding an extra line can help us (see Chapter 3).

So, what sort of changes might be useful?

- Moving points around (but preserving the initial scenario).
- Trying to make a special situation (for example, a line of symmetry).
- Moving points to extreme positions.

The following problems may be solved using more than one method; however, the worked solutions provided at the back of this book are based on the method introduced above.

Alina is asked to choose five different fractions and write them down in ascending order.

She thinks carefully and then writes:

$$\frac{1}{6} \qquad \frac{1}{5} \qquad \frac{1}{4} \qquad \frac{1}{3} \qquad \frac{1}{2}$$

a Suggest another set of fractions that she could have written.

b Suggest another set of fractions where none has a numerator of 1.

c Suggest another set of fractions, all of which have their numerator greater than their denominator.

Tip

There are lots of ways to do this. One way is to change the original set of fractions.

In a restaurant chairs can be arranged around square tables, as shown below.

  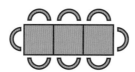

a How many chairs are needed if five tables are arranged in this way?

b Find the rule for the number of chairs needed for *n* tables.

The restaurant owner has a total of 55 chairs.

c Is it possible to use all 55 chairs with tables arranged like this? Give reasons for your answer.

The restaurant needs to be set up for 32 guests at a function. The party leader wants the guests to be split into two equal groups, so they will sit at two large tables.

d How many small square tables are to be used to make each of these large table settings?

Elliott has a bag of red, green and yellow coloured balls.

The probability of picking a red ball at random is $\frac{3}{7}$.

The probability of picking a yellow ball at random is the same as picking a green ball.

a There are nine red balls. How many balls are there in total?

Elliot adds more red and yellow balls to the bag. The probability of picking a green ball at random is now $\frac{3}{25}$.

b What is the total number of balls in Elliott's bag now?

53

Which is bigger?

a 400 g + 400 mg   or   0.5 kg – 90 g

b 0.1 km + 150 cm   or   110 m – 900 cm

c 0.75 hours + 600 seconds   or   50 minutes + 0.1 hours

On average there are over 100 000 strands of hair on a child's head. Blondes average about 140 000 strands, brunettes average 108 000 strands and redheads average 90 000 strands. Hair grows at a rate of about 150 mm a year. The average person loses roughly 0.25 per cent of their hair strands each year.

If the hair does not grow back:

a roughly what percentage of their hair strands would redheads lose after 82 years?

b how many years would it take for an average blonde to lose one tenth of their hair strands?

 Calculate the mean of the following three numbers:

$$\sqrt{24} \qquad \sqrt{54} \qquad \sqrt{96}$$

> **Tip**
>
> What change will you need to make before you can add the three numbers?

I choose three consecutive integers and add them together.

5 + 6 + 7 = 18

I notice that 18 is also the result of 6 × 3. I select another set of three consecutive numbers and add them together.

2 + 3 + 4 = 9

I notice that 9 is also the result of 3 × 3.

It looks like the sum of three consecutive numbers is the same as the middle number multiplied by three.

a Is this always true?

b Can you find a rule for adding four consecutive integers?

c How might you extend your rule to add five or more consecutive integers?

Animals in zoos are normally kept behind double fences to keep them in and the public out.

This style of circular pen is used for the chipmunks and rats. It has a diameter of 20 m. It is surrounded by a double fence with a gap of 2 m between the fences.

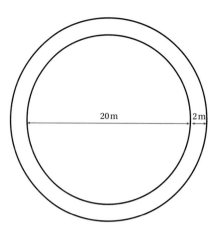

a How much longer is the outside fence than the inner fence?

b Suppose the gap was changed to 3 m. What would the difference be then in the length of the fences?

c What is the general rule?

d What happens if the pens are square or rectangular in shape?

 23.64 × 805 = 1903.02

How can you tell that this statement is incorrect?

Four snails have a race.

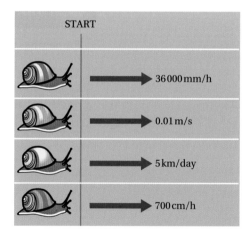

Assuming that the snails move at a constant speed to the finish line, which one will get there first?

55

The diagram shows two cereal packets.

 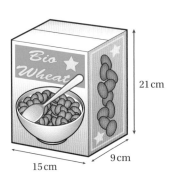

Tip

There are several steps to this problem. You could start by calculating the volume and surface area of each packet.

Mary says, "I think the Choc Flakes packet has the greater volume."

Kelly says, "I think the BioWheat packet has the greater surface area."

Raymon says, "Only one of you can be correct. Whichever packet has the greater volume must also have the greater surface area."

Who is correct? Give reasons for your answer.

Daisy draws a shape on a coordinate grid as shown below:

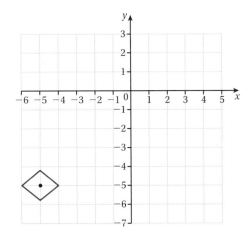

Tip

Remember to use vector notation to describe a particular translation.
Is there more than one solution to each part of this question? If so, can you write your answers in a more general way (using words or mathematical notation)?

She notices that the centre of her shape (marked with the black dot on the diagram) lies on the line $x = -5$.

a How could Daisy translate her shape so that its centre remains on the line $x = -5$?

Daisy's shape also lies on the line $y = x$.

b How could you translate Daisy's shape so that its centre remains on the line $y = x$?

c How could you translate Daisy's shape so that its centre lies on the line $y = x + 3$?

Mikey is asked to investigate what happens when a shape is reflected in two perpendicular mirror lines in turn.

He draws the following picture and makes the observation, "Ah! It's just the same as a translation of the original shape."

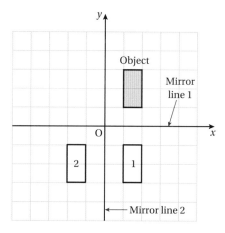

Mikey's teacher looks at his work and asks, "Are you sure?"

a Suggest what Mikey could do next in his investigation to address this

b By considering the starting point shown in the diagram below, decide whether or not you agree with Mikey's first idea.

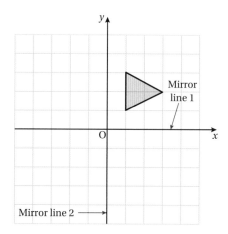

Melissa has designed this logo for her new business. It is based on an isosceles triangle with a semicircle on each side.

Calculate the perimeter of Melissa's logo. Give your answer correct to 2 dp.

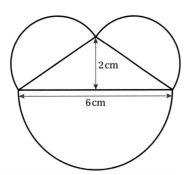

Diagram not to scale

> **Tip**
>
> Remember: You can use Pythagoras' theorem to calculate missing lengths in right-angled triangles.

 a What's the same and what's different about the following fractions and their decimal equivalents?

$$\frac{1}{9}, \frac{1}{99}, \frac{1}{999}, \frac{1}{9999}, \frac{1}{99999}, \frac{1}{999999}$$

b Say how you can use your findings to work out the equivalent fraction to each of these recurring decimals. Do this part without a calculator.

    i 0.7777... (repeating 7s)

    ii 5.3333... (repeating 3s)

    iii 0.14444... (repeating 4s)

    iv 2.5000900090009... (repeating 0009s)

> **Tip**
>
> You know about 0.1111... What will 0.2222... be?
> How can this help?
> Remember to consider the equivalent decimal number to each fraction.

Each of the seven small triangles in the diagram is isosceles. The large outer triangle is also isosceles.

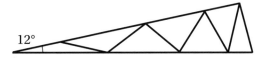

Diagram not to scale

12°

Given one angle, can you work out all of the others?

Jessie says, "No. You don't have enough information."

Helen says, "Yes. You can work out all of the other angles."

Pete says, "I wonder what would happen to the other angles if the first one was 10° instead of 12°."

a Decide whether Jessie or Helen is correct. Make sure you give reasons for your answer.

b Try out Pete's idea. What would happen if he started with an angle of 10°?

c Is there a different starting angle for which this *will* work?

Jack holds out a £1 coin and tries to cover up the Moon. He can't make it fit; the Moon is too big. He gets a friend to hold the coin steady and moves backwards until the coin is a good fit over the Moon. He finds that he has had to walk approximately 2.5 m away from the coin.

Given that the Moon has a diameter of 3474.8 km, and a £1 coin has a diameter of 22.5 mm, use Jack's findings to estimate the distance of the Moon from the Earth.

When you have done a lot of problem solving you might come across problems that are similar to others you have worked on before. Even if they are not identical you might find that some of the ideas you used in the earlier problem can help you out. Recognising a problem includes recognising which areas of maths will be helpful in solving the problem, as well as recognising similarities with a previous problem.

> Multiply two consecutive even numbers. Do you always get a multiple of 8? Give reasons for your answer.

You could start by trying a few out to see if it appears to be true (but this won't prove it if you think it is the case):

$$2 \times 4 = 8 \qquad 6 \times 8 = 48$$
$$4 \times 6 = 24 \qquad 10 \times 12 = 120$$

All of these are multiples of 8. It is looking good so far (but you haven't justified it).

How could you approach this problem? Think of maths topics that might be relevant. You could look at factors, but you could also use some algebra.

Try writing an algebraic expression for the problem. What can you use for the two numbers? If you call them a and b then that doesn't include the idea that they are **even** numbers. You could call them $2a$ and $2b$. You need a and b to be consecutive whole numbers. This tells us that either a or b is even. If you do $2a \times 2b$ you get $4ab$ but one of a and b is even so the whole thing must be a multiple of 8.

Here is a slightly more formal way to do the same thing, using just a single letter. If the first even number is $2a$ then you can call the next one $2a + 2$. When you multiply these you get $2a(2a + 2)$. This is equal to $4a^2 + 4a$, which is $4a(a + 1)$.

Let's look at $4a(a + 1)$.

If a is odd then $a + 1$ is even and we have $4 \times$ odd $\times$ even, which is a multiple of 8 (because 4 multiplied by an even number is a multiple of 8). If $a + 1$ is odd then a is even and we have $4 \times$ even $\times$ odd, which is a multiple of 8. So the whole thing is a multiple of 8.

> Choose any three consecutive positive integers. Add them together. Is the answer always a multiple of 3? Is the answer always a multiple of 6? Give reasons for your answer.

This problem is similar to the previous one, so you could use some similar techniques.

First, try a few numbers just to get a feel for what is going on. Then use some algebra and try to create an expression to describe what is happening.

$1 + 2 + 3 = 6$: this is a multiple of 3 and also a multiple of 6

$2 + 3 + 4 = 9$: this is a multiple of 3 but is not a multiple of 6

You have already shown that the answer is **not** always a multiple of 6 because you have found one that doesn't work.

Using algebra, you could call the first number a. The next number will therefore be $a + 1$ and the one after that will be $a + 2$. This is similar to what we did in the previous example so you are using the experience you gained doing that question.

When you add these together you get $a + a + 1 + a + 2$, which equals $3a + 3$. To show that this is clearly divisible by 3 you could factorise: $3(a + 1)$, to show that you have an integer $(a + 1)$ multiplied by 3.

If you call the middle number n then it happens to be even easier. The three numbers are: $n - 1$, n, $n + 1$, and adding these gives $3n$. Note that you didn't use three different letters to describe the numbers. This is often useful.

You can use some similar techniques to help you work on this next problem.

> Think of a number (a positive integer), square it and subtract the number you first thought of. When is the answer a prime number? Give reasons for your answer.

$$1^2 - 1 = 0 \qquad 3^2 - 3 = 6 \qquad 5^2 - 5 = 20$$
$$2^2 - 2 = 2 \qquad 4^2 - 4 = 12 \qquad 6^2 - 6 = 30$$

So far this looks interesting. You always seem to get an even number as the answer. Does that seem sensible? Will that ever change?

Only one of the answers is a prime number: $2^2 - 2 = 2$

There are several ways you could continue here, but one is to use some algebra like you did in problems 1 and 2.

Call the number you thought of n. Then you get $n^2 - n$.

This can be factorised to give $n(n - 1)$. But a prime number has exactly two factors, so if this is going to be prime then one of the numbers n and $n - 1$ has got to be equal to 1.

If $n = 1$ then you get $1^2 - 1 = 0$, which doesn't work (because 0 is not a prime number).

If $n - 1 = 1$ then $n = 2$ and you get the answer you know about already: $2^2 - 2 = 2$

All of the others will have extra factors, so they cannot be prime numbers.

For example: $7^2 - 7 = 7 \times 6$

The answer is a prime number only when the number first thought of is 2.

The following problems may be solved using more than one method; however, the worked solutions provided at the back of this book are based on the method introduced above.

The table shows data for the number of apples in a sample of 1 kg bags at two different supermarkets A and B.

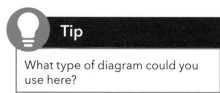
	A	B
Lower quartile	6	7
Median	9	8
Upper quartile	10	13

Which supermarket has the bigger apples? Justify your answer.

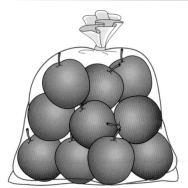

A farmer has started making crisps to sell at the weekly farmers' market. She does not want to buy expensive packing equipment, so she asks her husband and son to pack the crisps for her each week. They must make each packet a similar weight.

At certain times the farmer does a random test on some of the packets.

These were the results one day:

Husband: 57.8 g, 61.1 g, 62.3 g, 58.9 g, 59.5 g, 60.6 g, 60.1 g, 58.8 g, 58.5 g, 61.3 g, 59.5 g

Son: 58.1 g, 58.7 g, 59.3 g, 58.9 g, 59.3 g, 58.8 g, 60.4 g, 59.1 g, 59.4 g, 58.9 g, 60.2 g

The packets need to be labelled with the weight. The farmer plans to label them as being 60 g. Is she right to do this?

Give your reasons.

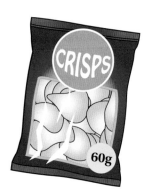

Prices for laying lawns are directly proportional to the area of the lawn needed.

For a lawn measuring 8 m by 10 m, there is a charge of £320.

For a lawn measuring 12 m by 12 m, there is a charge of £576.

a How much will it cost to lay a lawn measuring 12 m by 8 m?

b To lay a different lawn costs £560. If the measurements are in whole metres, what could the dimensions of the lawn be?

Nick has a poster on the wall in his office.

ALWAYS GIVE 100% AT WORK

12% MONDAY
23% TUESDAY
40% WEDNESDAY
20% THURSDAY
5% FRIDAY

Nick's poster

ALWAYS GIVE 100% AT WORK

14% MONDAY
18% TUESDAY
29% WEDNESDAY
36% THURSDAY
3% FRIDAY

Bernard's poster

Assuming that this poster refers to time rather than effort, answer the following questions.

a If Nick actually worked in this way, for how long would he work on a Friday?

b How much longer would Nick work for on a Wednesday compared to a Monday?

Bernard has a similar poster.

c Why can't you tell for certain who works for longer on a Tuesday?

Tip

You will need to decide how long a working day is. What is its usual length?

Cooking times for poultry are 50 minutes per kilogram plus an additional 30 minutes.

For Christmas lunch, Fergus needs to cook a 3 kg chicken and a 7.5 kg turkey.

He wants both birds to be ready at the same time and he needs 20 minutes to carve and prepare them once they are taken out of the oven. He wants to serve lunch at 1:00 pm on Christmas Day.

a How long will the turkey be in the oven by itself before being joined by the chicken?

b Find a formula to find the time taken, t minutes, to cook a chicken weighing m kg.

For New Year's Day Fergus plans to cook a goose. This goose will take 5 hours and 5 minutes to cook.

c How much does the goose weigh?

d If Fergus plans to have a late lunch at 2:30 pm on New Year's Day, what is the latest time he can put the goose in the oven?

The probability of pop star Karri K releasing a song on a Thursday is 0.62.

If her song is released on a Thursday, the probability of this song becoming a number one single in the charts is 0.48.

If her song is released on any other day, the probability of it becoming a number one single is only 0.18.

What is the probability that Karri K will have a number one hit with the next song she releases?

Tip

What kind of diagram would be useful here?

Joyce told her teacher that $y = x$ and $y = -x$ can be drawn as two straight lines that are perpendicular to each other.

The teacher asked Joyce what the gradient of each graph is.

a Complete Joyce's responses:

The gradient of $y = x$ is … .

The gradient of $y = -x$ is … .

The teacher asked Joyce to write down the equation of the line that has the same y-intercept as the graph $y = 2x - 3$, but which is perpendicular to it.

b Write the equation that Joyce should give for the perpendicular line.

The teacher asked Joyce to write a statement on the board using the correct terminology to describe perpendicular lines, but to make it simple for the other students to remember.

c Write a statement for Joyce to write up for the class.

Pauline and Emma live in Salford and Woburn Sands, respectively, two villages that are 18 km apart.

The girls plan to meet up on Saturday at Emma's house.

Pauline plans to walk to Woburn Sands, leaving home at 9:10. She can walk at a steady speed of 6 km per hour.

Emma's mum will be cycling to Salford and should pass Pauline on the way.

Emma's mum plans to leave Woburn Sands at 9:40 to cycle to Salford. For the first 16 km, she will be travelling at 24 km/h, but the remainder of the route is very steep and she thinks she will not arrive in Salford until 10:40.

a When will Pauline meet Emma's mum?

b How far will Pauline be from her house when they meet?

c How far will Emma's mum have travelled?

d What time is Pauline expecting to arrive at Emma's house?

e What is the speed Emma's mum was travelling up the steep hill?

f What distance apart are they at 10:20?

The graph shows the journey of a car and a moped.

The moped rider travelled from Edinburgh to Glasgow.

The car driver took the same route to Glasgow but returned to Edinburgh.

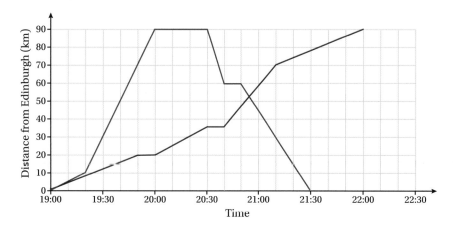

a How far did the moped rider travel in the first 25 minutes?

b How long did it take the car to travel the first 10 km?

c How many stops did the moped rider make and what was the total time she spent resting?

d What was the fastest stage of the journey for the car driver?

e What was the driver's speed at this fast stage?

f At approximately what time did the car and the moped pass each other?

g Roughly how far did the moped rider still have to go when she met the car driver?

h What was the average speed for the whole journey for the moped rider?

i What was the average speed for the return journey for the car driver?

Johnny says if you transform the graph f(x) = 2x + 4 by the transformation f(x) + 7, the intercept on the f(x) axis will become (0, 11).

Is Johnny correct?

Give reasons for your answer.

11

a Would you expect there to be any correlation between the average number of hours a secondary school student sleeps each night and the year group they are in?

b Write a hypothesis about this.

c Design a questionnaire to test your hypothesis.

d What type of diagram might be a useful way to display the data?

12

Two spiders are in a room that is 4.5 m long, 3 m wide and 2 m high.

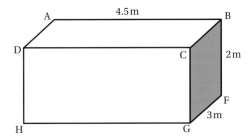

The male spider is at vertex A and the female spider is at vertex G.

a How far must the male spider crawl to get to the female if it only crawls along the edges of the room?

b If instead the male stays at A and the female spider travels along the diagonal of the face GCDH, then along edge DA to get to the male, how much shorter will her journey be?

13

Ryan draws a square of length 7.0 cm correct to the nearest millimetre. He then draws a circle that fits exactly inside the square.

Calculate the maximum difference between the area of the circle and the area of the square.

14

This sequence is generated by adding on the next even number each time.

0, 2, 6, 12, 20, 30 …

a Write down the next three terms.

b Find the nth term for the sequence.

c How does the algebra tell you that every term in the sequence must be even?

67

Cam keeps rabbits in his garden. They drink from two cylindrical water bottles, one small and one large.

The smaller bottle has a diameter of 8.5 cm and a height of 23 cm.

The larger bottle has a radius of 5.5 cm and a height of 25 cm.

How much **more** water does the larger bottle hold? Give your answer to the nearest millilitre (ml).

Tip

Can you work out the volume of a cylinder? What is the link between cubic centimetres (cm^3) and millilitres?

Robert is teaching his maths class about linear programming. He has a problem for them to solve, using small interconnecting bricks.

He gives each student eight red bricks and six yellow bricks and tells them to model a problem based on how many chairs and tables a company could make with a limited supply of materials.

Each table requires two red bricks and two yellow bricks. Each chair requires two red bricks and one yellow brick.

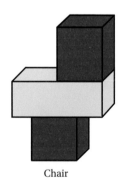

Chair

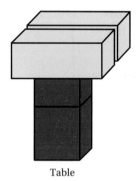

Table

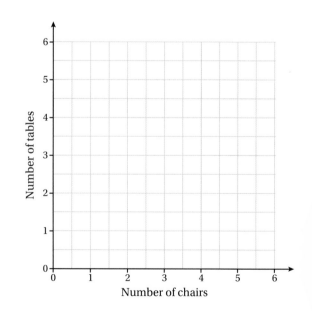

a Write an inequality to show the number of tables and chairs that could be made with the red bricks.

b Write an inequality to show the number of chairs and tables that could be made with the yellow bricks.

c Show these inequalities on a graph with axes as given.

The following question is posed to a class:

Find an equivalent expression to $10 \times \sqrt{15}$.

The teacher takes the answers suggested by the first four students:

A: $\sqrt{150}$ B: $\sqrt{1500}$ C: 40   D: 35

Which, if any, of their values is correct? Give reasons for your answer.

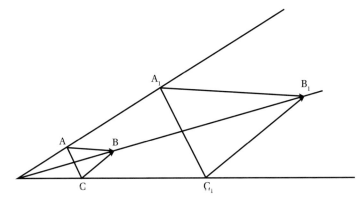

In the diagram above $\overrightarrow{AB} = 2\mathbf{a} - \mathbf{b}$, $\overrightarrow{CB} = \mathbf{a} + 2\mathbf{b}$ and $\overrightarrow{C_1B_1} = 3\mathbf{a} + 6\mathbf{b}$.

a Express $\overrightarrow{AC}$ in terms of $\mathbf{a}$ and $\mathbf{b}$.

b Express $\overrightarrow{A_1B_1}$ in terms of $\mathbf{a}$ and $\mathbf{b}$.

All standard television screens are mathematically similar. The size of a TV screen is advertised as the diagonal length, in inches.

Carlos owns a 36.7-inch wide-screen TV and measures its width as 32.2 inches.

a What is the aspect ratio (ratio of width to height) for a wide-screen TV?

Eddie is buying a 42-inch wide-screen TV (where 42 inches is the length of the diagonal).

b Work out the length and width of his new TV.

20

A square-based pyramid is made so that all of its edges are 2 m long.

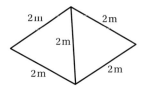

Tip

How can you make right-angled triangles? How could this help?

a How high is the pyramid?

b Suppose the edges were all 3 m long. How high would it be now?

c What is the general rule for a pyramid of side length a metres?

21

A pair of tongs has arms 25 cm long. When not in use, the angle between the two arms is 60°.

Alison uses the tongs to pick up a meatball 5 cm in diameter.

By how much will the angle between the two arms be reduced?

Tip

Draw a diagram to help you see what is happening. Which topic will you need to use?

22

The dial on Rupa's washing machine has a radius of 3.2 cm. It can be turned so that the arrow points to each of five equally spaced settings.

Rupa turns the dial clockwise from 0 through a curved arc of length 121 mm. Which setting has she chosen?

Tip

Think carefully about the sort of maths methods you need to use.

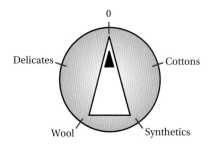

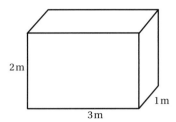

By how much could the volume of this cuboid vary if its side lengths were measured correct to:

a   the nearest metre?   b   3 sf?

c   the nearest millimetre?

Everything in Shop A is being sold at the same percentage discount.

Shop B is also selling all of its stock at a discount but it is not cutting its prices by as much.

A jacket now costs £209 in Shop A but £259 in Shop B. The recommended retail price (RRP) for the jacket is £310.

a   What percentage discount from the RRP is being offered in each shop?

A pair of trainers is priced at £62 in Shop A.

b   How much will they cost in Shop B?

 Consider each pair of expressions and decide which expression is the larger.

a $\dfrac{3}{\sqrt{5}}$ or $\dfrac{5}{2\sqrt{5}}$

b $\dfrac{4}{\sqrt{8}}$ or $\dfrac{25}{\sqrt{50}}$

c $\dfrac{10}{\sqrt{7}+3}$ or $\dfrac{6}{\sqrt{7}-3}$

Tip

How do you compare fractions? Can you apply the same technique here?

It is easy to get involved in working on a problem and to forget that you can sometimes use some common sense to tell roughly how big an answer ought to be. You will probably still need to carry out the calculations but this can give you a useful guide as to whether an answer is correct or not.

In A-level Biology there are several formulae that are used to calculate the actual size of something which has been viewed through a microscope. It is easy to make a mistake.

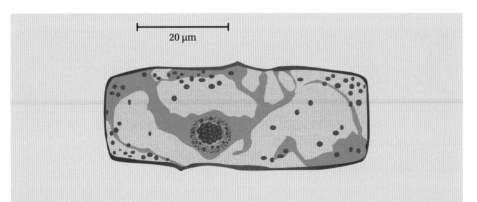

The diagram above shows a cell. The students have worked out the actual width of the cell using the formulae, but some of them have got the wrong answer. Which of these answers is correct? (You can work this out even you aren't studying biology.)

a 5.6 µm

b 56 µm

c 560 µm

d 5600 µm

The line above the diagram is 20 µm long and the cell is about three times as wide as that line, so it must be about 60 µm. The answer must therefore be **b** 56 µm.

A politician recently complained about the lack of depth of a particular policy. They described it as being 'a mile wide and an inch high'.

What is the area of a rectangle that is 'a mile wide and an inch high'? Give your answer in square metres.

You could begin by thinking about what a sensible sized answer would be. Would the area be equivalent to that of a tabletop, a classroom, school hall, playing field, or would it be even larger?

Jemal says: Malcolm says:

"If I started with a ribbon that was an inch wide and a mile long, and cut it into strips, I think it would cover the classroom floor."

"If I had a vacuum cleaner with a very thin nozzle (an inch wide!), how much carpet would I clean by the time I'd walked a mile? It takes about 20 minutes to walk one mile. If I was vacuuming for 20 minutes, I reckon I could cover about two rooms."

You could make use of other problem-solving techniques, such as drawing a diagram like the one below (note that it is not to scale!).

You will also need some knowledge about converting units (which might be given to you in an exam and which could be looked up). The conversions you need here are 1 mile = 1600 m and 1 inch = 2.5 cm.

To work out the area of the rectangle you need to multiply 1600 m by 2.5 cm, but you must use the same units for both measurements, so you can do 1600×0.025.

This will give the answer in m^2.

$1600\,m \times 0.025\,m = 40\,m^2$

A room that is 5 m by 8 m would have a floor area of $40\,m^2$ so it is about the size of a classroom, which does seem sensible.

The following problems may be solved using more than one method; however, the worked solutions provided at the back of this book are based on the method introduced above.

Two school basketball coaches want to do a study of each of their teams and need the average height of all the team members including the reserves.

Coach McKay used the data in the table on the right to find the mean height for his team.

Coach Cooksey used the data from the last health check on each member of his team to find the mean height. The data were as follows:

150 cm, 156 cm, 187 cm, 199 cm, 203 cm, 178 cm, 194 cm, 188 cm, 167 cm, 194 cm, 152 cm, 226 cm, 225 cm, 199 cm, 188 cm, 221 cm, 178 cm, 143 cm, 142 cm, 170 cm

a Which coach will have the more accurate mean height for his team? Give reasons for your answer.

b What is the mean height for the McKay team?

c Compare the mean heights and height ranges of the two teams.

Team McKay	
Height (cm)	Frequency
$145 \leqslant h < 155$	1
$155 \leqslant h < 165$	2
$165 \leqslant h < 175$	2
$175 \leqslant h < 185$	8
$185 \leqslant h < 195$	3
$195 \leqslant h < 225$	4

Running tracks come in two standard lengths. Indoor tracks are 200 m long and outdoor tracks are 400 m long.

a Simon's event is the 1500 m.
How many laps of each track will Simon run in his race?

b Denise will be running both the 5 km and 10 km race.
How many laps of each track will she run in each race?

The rule for calculating how long to cook a piece of lamb so that it is medium rare is '40 minutes per kilogram and an additional 25 minutes, at 180 °C'.

a Create a formula to calculate the cooking time for a piece of lamb of any size cooked this way.

Uncle Bertie wants to serve Sunday lunch at 12:30. His piece of lamb weighs 3.75 kg and it must stand for 10 minutes after it is taken out of the oven.

b Uncle Bertie thinks he needs to put the lamb in the oven at 9 am.
Is he right?

Hattie does some form of exercise each day.

Each day she decides at random to do one of the following:
swim 40 lengths, jog 10 km or complete a two-hour cycle ride.

a Draw a sample space diagram showing all the possibilities of the exercise Hattie could do on two consecutive days.

b What is the probability that she swims on both days?

c What is the probability that she does not do the same exercise on both days?

The ingredients for Fiona's recipe for a chocolate cake that serves eight people are shown below.

Cake	Icing
150 g dark chocolate	120 g dark chocolate
6 eggs	100 g hazelnuts
125 g unsalted butter	125 ml double cream
400 g chocolate spread	
100 g ground hazelnuts	

Fiona is having a dinner party and has invited 20 guests. She wants to have enough cake for each guest to have one slice.

a How much dark chocolate should Fiona buy?

b The dark chocolate is available in 200 g bars. How many bars will she need?

c Express the amount of chocolate used to chocolate left over as a ratio in its simplest form.

d On another occasion, Fiona reduces the recipe to serve six people. How much chocolate spread does Fiona need?

Five cleaners can clean an office building in two hours and 45 minutes.

a How long should it take three cleaners to clean the building? Give your answer correct to the nearest minute.

b Each cleaner is paid the same hourly rate, and their time is rounded up to the next whole hour. Does reducing the number of cleaners from five to three reduce the amount paid in wages?

The table shows the number of students in each year group at a new school.

Year 7	243
Year 8	176
Year 9	162
Year 10	88
Year 11	51

It has been decided there should be a student council of 20 students.

The head teacher wants to take a stratified sample of 20 students to be on this student council.

The deputy head wants to have four students from each year group on the student council.

a If the head uses her method, how many of each year group will be on the council?

b Give your opinion on the method you think is best.

Tip

What would be the problem with each method?

Two glasses that are the same size contain orange drinks.

The first glass is $\frac{1}{4}$ squash and $\frac{3}{4}$ water.

The second glass is $\frac{1}{6}$ squash and $\frac{5}{6}$ water.

The two glasses are poured into a large drinks bottle.

a What fraction of the resulting drink is squash?

Amit is serving drinks at a party. He has two large jugs of fruit punch. The jugs are identical in size, but one contains punch that is $\frac{2}{5}$ apple juice and the other contains punch that is $\frac{3}{10}$ apple juice.

b Amit wonders how the drink will change if he pours some from both jugs into a glass.

Is it possible for Amit to create a drink that is exactly $\frac{1}{2}$ apple juice?

A health campaign is targeting the high levels of sugar in low-fat foods.

One particular brand of low-fat yogurt is sold in 104 g pots and contains 5.9 g of sugar.

Another yogurt is sold in 90 g pots and contains 5 per cent sugar.

If the guideline daily amount (GDA) of sugar for an adult is approximately 90 g, how do the two yogurts compare?

Meg has 105 chocolate buttons. Grace has 95 chocolate buttons.

By what percentage will each girl need to change their number of chocolate buttons in order to have exactly 100 each?

Valerie told Doug the expression for the area of this rectangle is $x^2 - 7x + 6$.

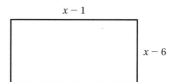

Doug said she was wrong. He told Valerie the expression for the area was $2(2x - 7)$.

a Who is correct? Give reasons for your answer.

Doug also (correctly) told Valerie the area was numerically 2 more than the perimeter. He said as an example that if the perimeter was 12 cm, then the area would be 14 cm².

b What are the dimensions of the rectangle?

Valerie did her calculations and told Doug that x could have two values. She said x could equal 2 cm or 9 cm. Doug burst out laughing and told her to use logic to work out why this was not true.

c Say why Doug found Valerie's statement funny.

Mr George has a box of board markers. It contains six blue markers, five black markers, seven green markers and two red markers. When he needs a marker, he picks one from the box at random, uses it and then places it back in the box.

On Monday Mr George needed a marker on two occasions.

a What is the probability that he randomly picked the same colour both times?

b What is the probability that he randomly picked a different colour the second time?

c If Mr George forgets to put the first marker back in the box, what is the probability now of taking the same colour of marker both times?

d Is the probability higher or lower when he does not put the marker back than when he does?

The cost of an extension to a house is directly proportional to the floor area of the extension. The floor area of an extension was 4 m by 5 m and its cost was £15 000.

a What would be the largest floor area possible for £23 000 if the dimensions must be in complete metres?

The extension could be completed in four weeks if the builder had three labourers working five-day weeks from 8 am until 5 pm with him.

b How many labourers are needed to finish the work within three weeks?

c With the builder and this number of labourers, exactly when after their start day should the job be completed?

The labourers are paid a daily rate, and the builder has decided that the jobs left on the last day are best done by him on his own.

d How many extra hours' work would the builder be doing without the other labourers, and will the entire job still be completed within three weeks?

The labourers are paid a flat rate of £100 per day and the total price charged by the builder covers labour, materials and profit in the ratio 26:39:25.

e What percentage of the price charged is profit?

Susanne wants some carpet fitted in her flat.

Two firms in her area specialise in floorings: Carpet Lay and Underfoot.

She has checked their websites and plotted their charges on a graph to compare their prices.

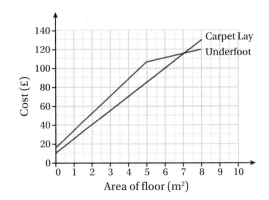

a How much does each company charge to lay 5 m² of carpet?

b Which company is the cheaper to fit 7.5 m² of carpet?

c What is the largest whole number of square metres that Carpet Lay would lay at a lower price than Underfoot?

Susanne needs a total of 9 m² of carpet laid.

d Which company should Susanne use?

e How much will this cost her?

The depth of water in Oban Bay depends on the time of day and the tides.

Time (hours am)	0	1	2	3	4	5	6	7	8	9	10	11	12
Depth (metres)	7.0	10.7	12.0	12.1	10.5	7.5	4.6	2.6	1.5	1.2	1.9	4.1	8.0

a Copy the axes, plot the points and join them up with a smooth curve.

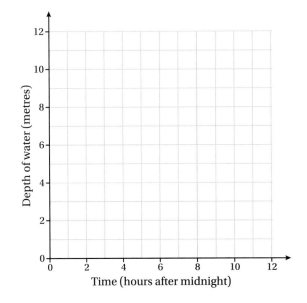

The Skye ferry needs at least 4 metres of water to stay afloat.

b Between what times should it not attempt to dock?

c Is there going to be another period of 'danger' time within a 24-hour day?

The data in the table shows the number of microbes present in a petri dish after certain periods of time.

Number of microbes	315	1215	90	447	123	690	63	378	258
Number of minutes	10	20	5	12	6	15	4	11	9

a Estimate how many microbes were present after eight minutes.

b Is it possible to predict how many microbes might be present after 50 minutes? Give reasons for your answer.

Peter is 2.03 m tall. He and his three shorter friends are planning on going camping. They will be using a traditional tent in the shape of a triangular prism.

What dimensions must their tent be if Peter is to be able to comfortably stand upright and lie flat to sleep alongside his three friends?

Tip

You will need to make some assumptions in order to answer this question. You know how tall Peter is but how much space will he need from side to side?

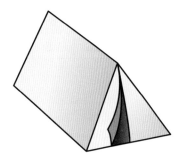

Mrs Grant decided to do a survey with her Year 7 class to find out how much time they spent playing games on their phones.

She recorded the average amount of time the students spent playing games each day.

a Copy and complete the histogram.

b Complete the frequency table.

Mrs Grant told the class that in her opinion anyone who spent more than 100 minutes per day playing games probably wasn't spending enough time on their homework.

c How many students were not doing enough homework?

Time (minutes)	Frequency
$10 < t < 20$	5
$20 < t < 60$	4
$60 < t < 120$	
$120 < t < 150$	3

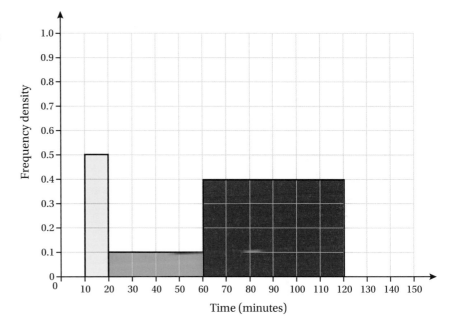

Sometimes it's useful to make a change to a diagram or scenario so that it is easier to see what's going on.

> The diagram shows a red square with a purple circle drawn through its vertices and a blue square drawn so it touches the purple circle on all sides.
>
>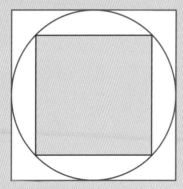
>
> What is the relationship between the area of the red square and the area of the blue square? Say how you know.

Before reading further, have a go at the question.

What ideas have you got?

Here are some things you can try.

- Get a rough idea of what the answer will be; you know that the blue square is bigger than the red square, but it is certainly less than, say, four times the area. This is **not** the way to work out the final answer, not least because you can't provide an explanation, but it will help you decide if your answer seems reasonable later.
- You don't know any measurements, so do you need to provide some?
- Will it help to call the side length of the red square x? Or the radius of the circle could be r. Maybe this will be useful and maybe it won't.
- Try putting on some extra lines. Will they help?

Tip

If you can't do it please don't worry. This is a mathematical problem that is supposed to be difficult. Remember, if you can just 'see' the answer then it wouldn't be a mathematical 'problem'.

Did you manage to answer the problem?

If not, here's one way that you could have solved it by changing the way you look at the diagram.

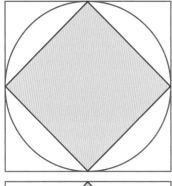

If you rotate the red square nothing vital changes.

The area of the red square is still the same and the purple and blue shapes are unaffected.

Does this help you to see the relationship between the red square and the blue square?

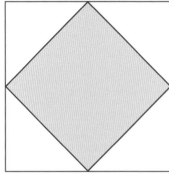

To make this very clear you can remove the circle.

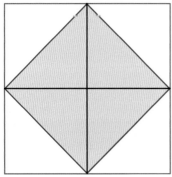

Then you can divide the shape up like this.

Tip

The extra lines are helpful after you've rotated the red square.

Tip

You had a rough idea that the area of the blue square is between one and four times the area of the red square, and it is. Estimating the answer can be a useful check that your final answer is sensible.

Now you can say that the area of the red square is half the area of the blue square.

Here is an example of a full answer.

If you rotate the red square so the vertices of the red square meet the middle of the sides of the blue square you do not change its area.

The two extra lines I have added join the vertices of the red square, cutting it into quarters that are right-angled triangles. These lines also join the middle of the sides of the blue square, cuttting it into quarters that are squares.

The total area of the red right-angled triangles is half the area of the blue square.

So the area of the red square is half the area of the blue square.

Tip

You may have answered this question by working out some measurements in terms of *x* or *r* instead of using this method. This is also a good way to solve the problem.

Here is another question.

The formula for a triangular number is $T = \frac{1}{2}n(n+1)$.

Prove that $8T + 1$ is always a square number.

This can be done algebraically: $8T + 1 = 8 \times \frac{1}{2}n(n+1) + 1$

$$= 4n^2 + 4n + 1$$

This factorises to give $(2n + 1)^2$, which is square.

An alternative method is to use a diagram.

There are clearly 8 copies of a triangular number, there is 1 in the middle and it makes a square. To complete this explanation you need to show that it will work for all triangular numbers.

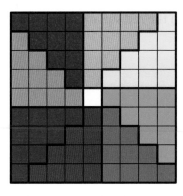

The following problems may be solved using more than one method; however, the worked solutions provided at the back of this book are based on the method introduced above.

A positive number is four more than five times its reciprocal.

a What is the number?

Another positive number is five more than six times its reciprocal.

b What is this number?

c What is $(n - 1)$ more than n times its reciprocal? Say how you know.

Tip

The reciprocal of 2 is $\frac{1}{2}$.

Here are six views of the same cube.

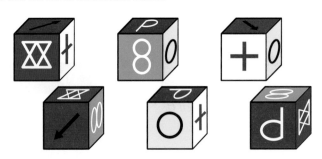

Which symbol is opposite the red cross?

Tip

Imagine holding the cube in different positions. Try to work out which faces are opposite each other.

Would you rather:

a   receive 80% of £10 or 75% of £12?

b   have 50% of one chocolate cake or 25% of two chocolate cakes?

c   buy a CD in store for £12.99 or buy it online with a 5% discount code but a postage charge of £1.99?

Sections of decorative fencing are made with metal and wooden rods.

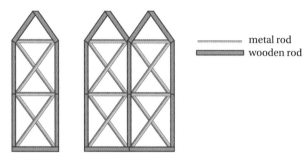

a   How many metal rods will be needed in the next pattern?

b   How many wooden rods are needed for the fourth pattern?

c   Find the number of metal rods needed for the nth pattern.

The metal rods cost £2 each and the wooden rods cost £2.50 each.

d   Sue needs a fence consisting of 12 sections. How much will this cost?

Yasmin is looking at a 3-D solid. If she looks at it directly from the side this is what she sees:

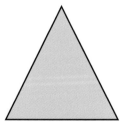

a   If Yasmin now looks at the solid from directly above, what could she see?

b   How many different answers are there?

Describe the shapes as accurately as possible.

Simon is a landscape gardener and he needs to hire a small digger for his next job.

The cost of hiring a digger from JCEs is shown as a straight-line graph.

MoveIt Builders also hire out diggers but they have a basic fixed charge in addition to their daily charge. Their charges are also shown as a straight-line graph.

a Which graph line refers to the charges for which company?

b What is the basic fixed charge made by one of the companies?

c How much per day does JCEs charge?

d Is MoveIt Builders' daily charge more or less than what JCEs charge? By how much?

e What would be the total cost if Simon hired the digger from MoveIt Builders for five days?

f What would be the total cost if Simon needed the digger for two weeks and he hired it from MoveIt Builders?

g If Simon needed the digger for two weeks, would it be cheaper from MoveIt Builders or JCEs?

h For the two-week period, how much less would it cost to hire from the cheaper company?

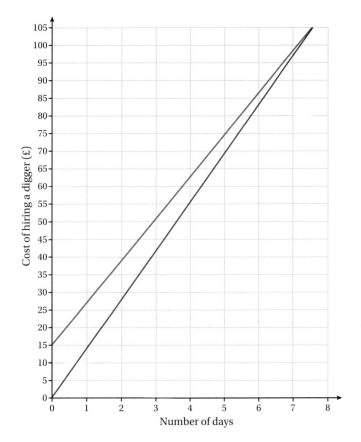

Jenny is given a set of instructions to create an image:

Mark a point on your page and label it 'O'.

Draw a 1 cm square with its bottom left-hand corner 1 cm to the right of O.

Enlarge this square by a scale factor of 2 about O.

Translate the smaller square 1 cm to the left.

Rotate the two smaller squares 90° clockwise about the centre of the larger square.

Rotate the image (up to and including the fifth step) 180° about the centre of the larger square.

a What should Jenny's completed image look like?

b How many lines of symmetry does the resulting image have?

c What order of rotational symmetry does the resulting image have?

d Design a set of instructions to create the image below.

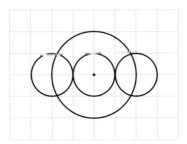

Arthur has just bought a circular dining table of diameter 240 cm. It comes flat-packed to make it easier to transport. His front door is a rectangle, 2.1 m tall and 1.05 m wide. Will he be able to fit the tabletop through the door?

 Tip

It should be clear that the circular table is not going to fit through the door if held vertically or horizontally. Arthur will need to tilt it to try to get it inside.

Susan is y years old. Mark is z years old.

Together their ages total 20 years.

Susan is 8 years older than Mark.

How old are Susan and Mark?

The number of living bacteria cells on the petri dish in the science lab is inversely proportional to the number of hours it remains in the refrigerator.

The number of living bacteria cells on a petri dish that has been in the refrigerator for four hours is 20 000.

Jamila thinks that the longer the time in the refrigerator, the smaller the number of bacteria cells. She worked out the number of bacteria cells she thought would be on the dish.

a Find a formula Jamila could use.

b What number of bacteria cells would Jamila's formula give for a dish that has been in the refrigerator for eight hours?

c What problems are there with Jamila's formula?

Sarah draws the graph of $f(x) = (x - 2)(x + 2)$.

Sarah decided that if she made the transformation $f(x) + 4$ on her graph she would be sketching the graph of $f(x) = x^2$.

Is Sarah correct?

Give reasons for your answer.

 Compare each pair of numbers and decide which symbol out of <, > or = should be placed in between them.

a $\dfrac{2}{3}$ 0.66   **b** 0.25 $\dfrac{8}{32}$

c $\dfrac{3}{11}$ 0.273   **d** 0.05 $\dfrac{1}{22}$

e 0.5   0.4999...

Eric is reorganising the hamsters in his pet shop.

If he puts three in each cage, there is one hamster left with no cage to go into.

If he puts four in each cage, one cage is left empty.

a How many hamsters does Eric have?

b How many cages does he have?

a Find three consecutive even numbers that add up to 228.

b Find three consecutive odd numbers that add up to 291.

c The sum of the squares of two consecutive numbers is 1301. Find the numbers.

d The difference between two numbers is 6. The sum of the two numbers is 80. Find the numbers.

What conditions are there on the values of x and y if a can be anything and each equation is always correct?

a $(a^x)^y = 1$

b $(a^x)^y = a$

c $(a^x)^y = \sqrt{a}$ $(a \geqslant 0)$

The planet Mars takes 24.6 hours to rotate on its axis. The Earth takes 24 hours to rotate on its axis.

a Express these times in hours as ratios in their simplest form Mars : Earth, in integer form.

b Express these times in minutes as ratios in their simplest form.

The mass of Mars is 6.4×10^{23} kg. The mass of Earth is 6.0×10^{24} kg.

c Express the masses as ratios.

The highest mountain in the Solar System is Olympus Mons, measuring approximately 27 km high. Olympus Mons is on Mars.

The highest mountain on Earth, Mount Everest, is 8848 metres high.

d Using a suitable degree of accuracy, express the heights as ratios: Olympus Mons : Mount Everest

The area of a square is given as $4x^2 - 12x + 9$.

Find an expression for the perimeter of the square.

a Create a formula, in terms of x, to calculate the area of the shaded border.

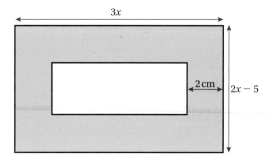

b The area of the border is 204 cm^2. What are the dimensions of the outer rectangle?

The sequence 2, 10, 24, 44, 70, ... can be written as
$(1 \times 2), (2 \times 5), (3 \times 8), (4 \times 11), (5 \times 14), ...$

a Write an expression for the nth term of the sequence.

One of the terms in the sequence is 234.

b Write this term in the form $(a \times b)$, where a and b are values to be worked out.

Anna is planning to do something silly to raise money for charity.

Will it be cheaper for her to take a bath in cola or to take a bath in baked beans?

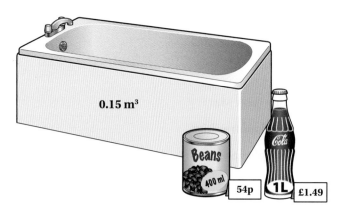

21

100 people were asked to complete a questionnaire for the RSPCA about their pets.

Out of the 100 people, 17 had both a cat and a dog. A total of 54 had either a cat or a dog, but not both. 30 people owned a dog.

a If one person was chosen at random, what is the probability they had neither a cat nor a dog?

b If one person was chosen at random, what is the probability they owned only a cat?

22

Aaron went to the local fruit market and bought four apples and five bananas for £2.05.

Peter went to the same fruit market and bought three apples and seven bananas for £2.35.

Aaron and Peter asked their maths teacher for an easy method to find the price of an apple and the price of a banana.

She told them they could create an equation for the cost of the fruit each of them bought and solve them simultaneously.

a Write the two equations.

b What is the cost of an apple?

c What is the price of a banana?

d Can you show this same information as straight-line graphs? How would you use your graphs to find the price of the apple and the banana?

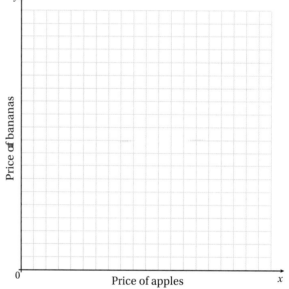

23

In how many ways can you get from flag A to flag B:

a using a single transformation?

b using exactly two transformations?

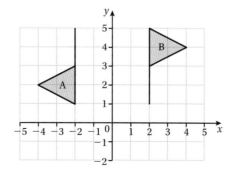

Tip

Remember to provide all of the information necessary to fully describe each transformation.

24

Triangles ABC and XYZ are similar shapes.

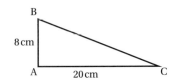 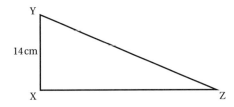

In triangle ABC, length AB is 8 cm and AC is 20 cm.

In triangle XYZ, length XY is 14 cm.

a Work out the area of triangle XYZ.

b Express area of triangle ABC : area of triangle XYZ in its simplest form.

c What type of numbers are in the ratio? Can you say why?

25

a Copy and complete the table of values for the graph
f(x) = x^2 – 6x – 5.

x	–2	–1	0	1	2	3	4	5	6
f(x)									

b Write down the coordinates of the turning point of the function f(x).

c Write down the exact values of the roots of the function f(x).

d Write down the coordinates of the turning point of the function –f(x).

e Write down the exact values of the roots of the function –f(x).

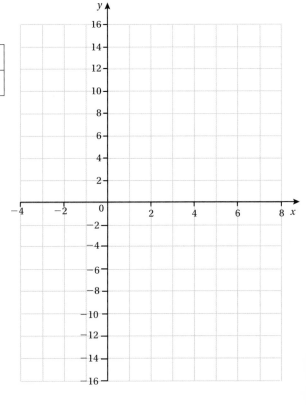

26

Zain wanted to carry out the transformation f(x – 8) on the function
f(x) = x^3. He wrongly used the translation vector $\begin{pmatrix} 0 \\ -8 \end{pmatrix}$.

a Say why this is wrong.

b Starting from Zain's new, incorrect version, what translation vector
should he use to get to the correct answer?

Brandon is learning about the sine and cosine graphs.

Brandon's teacher asked him to sketch both the sine and cosine graphs, showing the coordinates of the points where the graphs cross the axes.

a Sketch the sine graph and the cosine graph on the same set of axes.

Brandon's teacher then asked him to describe the sine graph in terms of a transformation of the graph of $f(x) = \cos(x)$.

b Write down this transformation.

The teacher then asked Brandon to change it around and to describe the cosine graph in terms of a transformation of the sine graph.

c Write down this transformation.

d Would you describe these transformations as reflections or translations?

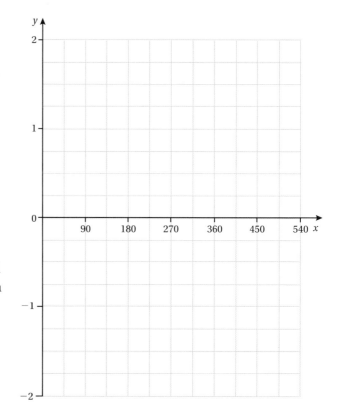

A square is surrounded by four identical rectangular tiles, as shown.

a Find an expression for the total area of the four identical tiles.

b Find an expression for the area of the large square.

c What is the area of the unshaded square?

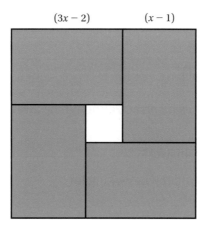

93

In an exam, students are asked to calculate the area of the shape shown below. They are instructed to give their answer correct to 1 dp.

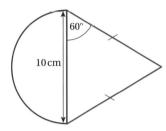

Ollie's answer is shown here:

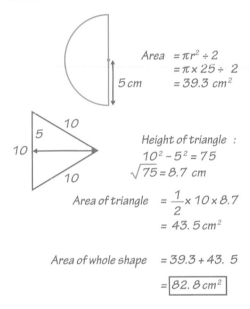

Area $= \pi r^2 \div 2$
$= \pi \times 25 \div 2$
$= 39.3 \ cm^2$

Height of triangle :
$10^2 - 5^2 = 75$
$\sqrt{75} = 8.7 \ cm$

Area of triangle $= \dfrac{1}{2} \times 10 \times 8.7$
$= 43.5 \ cm^2$

Area of whole shape $= 39.3 + 43.5$
$= \boxed{82.8 \ cm^2}$

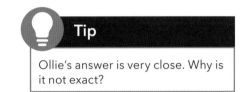

Tip

Ollie's answer is very close. Why is it not exact?

The examiner marks Ollie's answer as incorrect, although he awards some marks for Ollie's method.

a What is the problem with Ollie's answer?

b What is the answer that the examiner was looking for?

A square piece of paper, of side length 8 cm, is folded so that the bottom right-hand corner, C, meets the midpoint of the top edge AB.

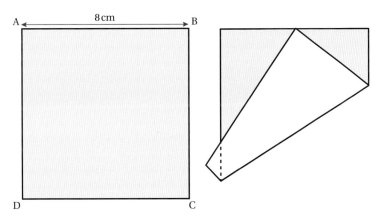

a Show that the three triangles created are mathematically similar.

b Show that the lengths of the sides of the smallest triangle are exactly a quarter the size of the largest triangle.

 a Use the diagram of an equilateral triangle to find the exact value of:

i tan 60°   **ii** sin 60°   **iii** cos 60°.

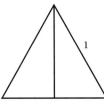

b How does this diagram show that sin 60° = cos 30°, and that sin 30° = cos 60°?

In some problems you need to make sure you get every possible answer, without missing any or getting the same answer twice.

Here is the flag of South Korea.

Around the edge of the central circle there are four 'trigrams'. These are made up of three lines. The one on the right has three solid lines.

And this one has a broken line at the bottom and the top, with a solid line in the middle.

If you just use solid lines and broken lines, what fraction of all the possible trigrams (made using three lines) appear on the South Korean flag?

First of all, do you understand the problem? There are other trigrams that haven't yet been drawn, such as the one on the right.

It looks as if you will need to work out how many possibilities there are. If you just start writing them down (which is a reasonable starting point) there is a risk that you will miss some of them, or maybe write the same one twice.

These are the ones you have got so far.

A better way to solve the problem would be to work systematically. You could start with three solid lines and no broken lines. That is easy because there is only one of them, as shown on the right.

Next you could have two solid lines and one broken line.

This example is systematic in the way the broken line has moved downwards. Can you finish off the rest of the trigrams?

The final list that you get is as follows.

There are eight of them and you can be confident that because you have worked systematically you haven't missed any out, or got any of them twice. Go back to the original question: **what fraction of all the possible trigrams (made using three lines) appear on the South Korean flag?**

There are four of them on the flag and you know there are eight altogether, so $\frac{1}{2}$ of the trigrams appear on the flag. (Rather neatly, they have chosen to use the four that look the same both ways up!) If you want to test out your ability to work systematically you could try to find all of the ways to make a diagram that uses four lines.

Sometimes you can work systematically but save yourself some time. Here's an example.

> The postman knew it was Cara's birthday because of the large number of cards and packages he had to deliver. When he asked Cara how old she was, she told him that yesterday her age was a square number but today it is a prime number. Can the postman work out how old Cara is?

You need to find two consecutive numbers where the first one is a square number (yesterday's age) and the next number is a prime number (Cara's age today, on her birthday).

Guessing numbers at random doesn't seem sensible. You could either list the prime numbers and find out whether the number before them is a square number, or you could do it the other way round and list all of the square numbers and see whether the next number is prime.

Both methods will work, but you should do the second one because working out some square numbers is much easier than listing prime numbers.

So far this is looking good. You are working systematically through the numbers and haven't missed any out.

You might have realised that you don't need to bother trying the odd square numbers from now on (so the rows 7 and 9 in the next table have been shaded out), because when you add 1 it will be even and can't be prime.

Which numbers work? Cara could be 2, 5, 17, 37 or 101 years old. Cara is an adult, so that means she is either 37 or 101. Hopefully it would be obvious from Cara's appearance whether she is 37 or 101, so it is likely that the postman **can** work out how old she is.

The following problems may be solved using more than one method; however, the worked solutions provided at the back of this book are based on the method introduced above.

n	Square number (n^2)	Add one to this	Is it prime?
1	1	2	Yes
2	4	5	Yes
3	9	10	No
4	16	17	Yes
5	25	26	No
6	36		
7	49		
8	64		
9	81		
10	100		

n	Square number (n^2)	Add one to this	Is it prime?
1	1	2	Yes
2	4	5	Yes
3	9	10	No
4	16	17	Yes
5	25	26	No
6	36	37	Yes
7	49		
8	64	65	No
9	81		
10	100	101	Yes

The first two questions on a worksheet are:

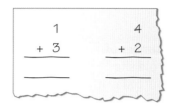

a Using positive integers less than 10, how many different addition questions could be on the rest of the worksheet? (Note that the order of addition is not important, so 1 + 3 is the same as 3 + 1, therefore these do not count as different questions.)

b If the worksheet includes negative integers greater than –10, how many different answers to addition calculations could there be?

 Insert the operators +, – and × into the three boxes below. You should use all of them once each.

5 ☐ 2 ☐ 10 ☐ –3

In how many ways can you get a negative result?

For a set of ten numbers the following information is known:

Mode = 3, range = 10, median = 3, mean = 6.

The numbers are all integers and there are only three different integers involved.

a Can you find the ten numbers?

For another set of four odd numbers the mean is 20, the mode is 3 and the median is 3.

b What are the numbers?

In this second set of four numbers, some changes are made, but the numbers are still all odd, the mode is still 3 but the median is now 5.

c What numbers are now in this set?

The numbers are changed again, so the four odd numbers have a mode of 3 and a median of 10.

d What are the numbers in this set now?

e What happens if the mode of four odd numbers is 3 and the median is 15?

f Investigate all possible sets of four odd numbers with a mode of 3 and mean of 20.

Tip

Working systematically can help you spot a pattern here. Remember: in this question the median must always remain an integer.

You have the fractions $\frac{2}{3}$ and $\frac{1}{5}$ and can combine them using any of +, –, × and ÷.

a What is the biggest answer you can get?

b What is the smallest possible answer?

$a + b = 1$

How many different pairs of values of a and b will satisfy this equation if:

Tip

Write out the answers in a systematic way.

a a and b are non-negative integers

b a and b are positive decimal numbers with 1 dp

c a and b are positive decimal numbers with 2 dp?

d How would your answer to part **a** change if a and b were allowed to take negative values?

Martin helps himself to sandwiches from a buffet lunch. He notices that the sandwiches have all been cut into identical triangles.

Tip

Think of a logical way of trying out all of the possible arrangements of the sandwiches.
You also need to decide on a sensible way to record your findings so that you can be sure you have found all of the possible polygons.

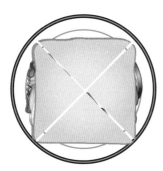

Martin places his sandwiches so that they fit together to create a square. He wonders how many different polygons he can create by arranging two or more sandwiches on his plate.

a Assuming that Martin matches the edges of the sandwiches, how many different polygons can he create, using:

i exactly two sandwiches

ii exactly three sandwiches

iii all four sandwiches?

b Is it possible for Martin to create all of the special types of quadrilateral using two or more sandwiches from his plate?

A factory makes packing boxes for cylinders.

Each box is a cuboid measuring 40 cm × 40 cm with a height of 15 cm.

a What are the maximum dimensions (radius and height) of the cylinders that can fit into each box?

b Make a guess as to which box has the most unused space.

c Use calculations to check your guess. What did you find?

> **Tip**
>
> The volume of a cylinder is $\pi \times r^2 \times h$.

> **Tip**
>
> The strategy from Chapter 8 could also be helpful here. You can ignore the height of the cylinders because it is the same for all of them.

Shona has five cards from a pack of playing cards: two jacks, a queen, a king and an ace. She shuffles the cards and picks two at random.

a Make a list of all the possible pairs of cards she could pick.

b What is the probability that she picks the two jacks?

c What is the probability that she does not pick any jacks?

 Two numbers are called co-prime if their highest common factor (HCF) is 1. For example, 9 and 10 are co-prime, but 10 and 14 are not (they have an HCF of 2).

You choose two numbers from 1 to 12 where the order is not important and you can have the same number twice.

a There are 78 such pairs. Show how you know this is correct.

b What fraction of the integer pairs from 1 to 12 are co-prime?

10

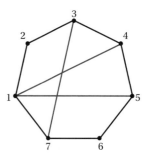

Jeff draws a heptagon and numbers the vertices in order. He then draws a diagonal on the shape and adds the numbers at the ends of the diagonal. He then repeats the process with another diagonal , then another, adding on the score each time. So far Jeff has a total of 21 (5 + 10 + 6).

Jeff draws all of the possible diagonals and adds the numbers.

a What is his final total?

Phil does the same thing with a hexagon.

Ali does the same thing with a pentagon.

b If Phil and Ali add their totals together, will they have a bigger number than Jeff?

💡 **Tip**

A diagonal of a shape joins two vertices that are not next to each other.

  11

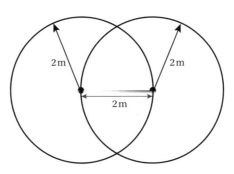

A splash pool is to be made from two interlocking circular holes, each of diameter 4 m.

The centres of the circles are 2 m apart.

The cost of the pool is broken down into the base (which is £800 per square metre) and the sides (£225 per metre of edge).

Work out the cost of the pool.

   12

Find values for a, b and c that make this equation true:

$245 = 1^a + 2^b + 3^c$

  13

Calculate the unknown length in the diagram below.

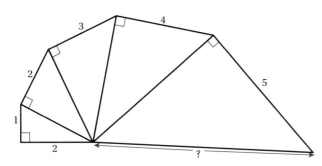

💡 **Tip**

Work with one triangle at a time. Why is the working easier if you leave your calculations in surd form?

To make an ordinary cake you mix the ingredients together and then bake the mixture.

A friendship cake mixture is different. You add a little extra milk each day and only bake some of the mixture.

Molly starts a new friendship cake mixture.

After a week, Molly divides her mixture into four, keeps one portion for herself and gives the other three portions to friends.

Her friends then keep their mixture for a week, adding milk and stirring it each day.

After a week, everyone divides their mixture into four portions, keeps one of them and gives the other three portions to other friends.

This process can continue for weeks.

After one week, Molly has given some of the original mixture away, so four people have some mixture.

a How many people have the mixture after four weeks?

b How many will have the mixture after six weeks?

c The population of the UK is about 66 million. How many weeks would it take until everyone could have received Molly's friendship cake mixture?

d Why is it not likely that everyone would receive it?

 The equivalent decimal of every fraction is either terminating (it stops) or recurring (it repeats forever).

How many of the fractions $\frac{1}{N}$, where N can take the value of any positive integer from 1 to 10, are recurring when you find their decimal equivalents?

Tip

Remember that a fraction represents a division calculation. You could use short division or another written method to turn each fraction into a decimal. Make sure you continue for long enough to decide whether the fraction represents a terminating or recurring decimal.

10 If you don't know what to do, do something

There is a strange scenario that exists if you are having difficulties with a problem. You could call it 'the problem with problem solving' as shown in the diagram.

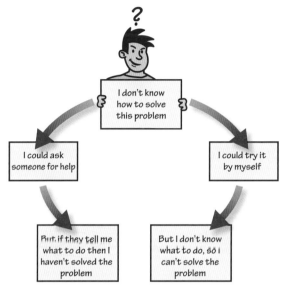

If you find yourself in this situation, one strategy you could try is: 'If you don't know what to do, do something.'

But make sure the 'something' you do is useful.

Some things that are **not** useful might include colouring a picture (unless it is a problem about colouring), or writing the question out in perfect handwriting. These are just things that keep you busy, but which are unlikely to help.

The previous chapters in this book outlined different problem-solving strategies. If you don't know what to do for a question, try one of these strategies. If the first one you try doesn't work, try a different strategy.

You might find it helpful to write down the things you do know (for example the key ideas from the question) and which areas of mathematics these relate to.

> Ben bought a bag of rice at the supermarket. He wonders how many grains of rice there are in the bag. Can you help him?

There are all sorts of things that seem to make this question impossible to answer.

- Not every bag of rice has exactly the same number of grains of rice in it.
- Different varieties of rice are different sizes (for example long grain rice is longer than risotto rice).
- Individual grains vary in size.
- You don't know how big the bag is. If he has bought a 1 kg bag then there will be approximately twice as many grains as there will be in a 500 g bag.

You will have realised by now that there isn't going to be an exact answer, so you need to work out a rough answer. It is often useful to work out a ball-park figure, and as long as it is of the right order of magnitude (which means it has roughly the correct number of digits) then that will be reasonable.

If you decide a problem is difficult and do nothing then it won't get solved! You could start to try some things that might be useful.

Count every grain of rice in a bag.

Nice idea, but you haven't got a bag of rice and the answer is probably rather big, so it will take a long time and it involves counting rather than doing maths, so let's try something else.

Weigh a single grain and then work out how many grains there will be in 1 kg.

This is nicer, no counting involved! A 2p coin weighs 7 g, so a single grain of rice will be extremely light, and certainly less than 1 g. Your scales aren't accurate enough to do this though (and you don't have any rice to weigh anyway).

Find out how many grains weigh 1 g and then multiply by 1000.

This is useful because it uses the fact that 1000 g = 1 kg. But you still don't have any rice to weigh.

Find out how big a grain of rice is.

Can you draw a picture of an uncooked grain of rice? (Uncooked rice is smaller than the fluffy, cooked rice.) This drawing is about half a cm long and maybe 2 mm high.

What shape is a grain of rice?

It's a bit like a cylinder but with sloping ends. In fact, it is roughly a cuboid! You could work out the rough volume of the grain of rice by assuming it is a cuboid that is $\frac{1}{2}$ cm by 2 mm by 2 mm.

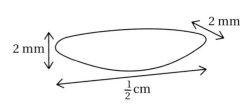

Work out the volume of a bag of rice and compare this to the volume of a grain of rice. In the kitchen cupboard at home you are probably used to seeing 1 kg bags of food (sugar, flour, pasta, rice, and so on). You can get a mental image of a 1 kg bag of sugar, using your ruler to help visualise the size, which is about 10 cm by 15 cm by 8 cm.

But the units are all different. That is awkward, so you could work in mm.

The volume of a single grain of rice is about 5 mm × 2 mm × 2 mm. This is 20 mm³.

The volume of the bag is about 100 mm × 150 mm × 80 mm. This is 1 200 000 mm³.

1 200 000 ÷ 20 = 60 000, so I estimate there are roughly 60 000 grains in a bag.

Ask yourself, does this seem reasonable?

If 60 000 grains of rice weigh 1 kg then about 60 of them would weigh 1 g. So this does seem possible. The errors in your estimate of the sizes are likely to be outweighed by the fact that rice grains are not cuboids, but they also don't fit together perfectly in the bag.

To start with we had no idea what to do. We tried out some different ideas and eventually worked out an answer.

The following problems may be solved using more than one method; however, the worked solutions provided at the back of this book are based on the method introduced above.

Jyoti and Indira are making towers with cubes.

Between them they have 12 cubes, each of different sizes.

There is a 1 cm cube, a 2 cm cube, a 3 cm cube, a 4 cm cube and so on up to the biggest cube, which measures 12 cm along each edge.

For the first towers they make, Jyoti uses the six smallest cubes and Indira uses the six largest.

a How much taller was Indira's tower compared to Jyoti's?

The girls then decide to make two towers of the same height using six cubes each.

b Work out a combination of cubes that each girl could have.

For a set of eight integers the following information is known:

The biggest number is 16.

Mean = 7.5. Mode = 3 and 5. Range = 15.

a Can you find the other numbers in the set?

b How many different solutions are there?

Tip

You know how many numbers there are: draw boxes for them and then write in what you know.

In a set of five positive integers the range is 15, the mode is 2 and the median is 5.

a If the mean is 7, what are the numbers?

b If the range, mode and median stay the same, what is the biggest possible mean value if this is an integer value?

c If the range, mode and median stay the same, what is the smallest possible mean value?

 Which of the statements about the number 12 345 678 912 are true?

i It is odd.

ii It is a multiple of 3.

iii It is a prime number.

iv Dividing the number by 5 leaves a remainder of 2.

v It is a multiple of 4.

Grandad Jones is twice as old as his son Geoff. Geoff was 29 years old when his own son Paul was born.

The total of the ages of the three family members is 131 years.

Find the ages of Grandad, Geoff and Paul.

The sum of the ages of three family members is 134 years.

Raj is five times as old as his daughter Savita.

His wife Lata is nine years Raj's junior.

How old is Lata?

For every eight students going on a school trip there must be an accompanying teacher.

a A school trip is planned for 140 students to go to the London Dungeons. How many teachers will need to be on this trip?

b Some extra teachers have decided they want to join the trip. There are now 21 teachers available. How many extra students could go on the trip?

c It was too late to take extra students, so the original 140 students went with the extra teachers. What is the new ratio of students to teachers?

The school hired some 42-seater coaches to transport everyone to London.

The drivers also went into the Dungeons and so there are more adults on the trip.

d Including the drivers, what is the new ratio of students to adults?

e What fraction of the group are now adults?

Two identical triangles are drawn using two diameters and two vertical lines, inside a circle of radius 7 cm.

If this diagram is placed on a centimetre-square coordinate grid, so that the centre of the circle is located at the origin, what will be the coordinates of points A and B?

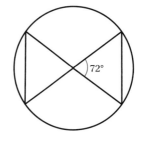

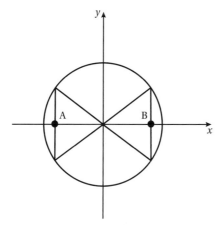

The manager of the local cinema did not want to give his employees a pay rise. He said on average they were earning more than the weekly national average wage, which is £130.

a Why would finding the median wage not be helpful for the manager?

Weekly earning in £	Frequency
61–80	2
81–100	5
101–150	9
151–200	2
201–250	1
251–300	1

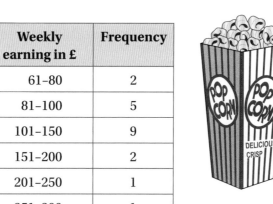

Feeling under pressure, the manager decided to increase the weekly wages of his two lowest-paid employees to put them in the £81–100 wage bracket.

b How have these pay rises affected the average wage of the employees? What has happened?

A magic square has numbers in each row, column and diagonal that add up to the same total (the 'magic number').

For the magic square below, the magic number is 2.1.

0.74		
$\frac{129}{150}$		
		$\frac{33}{50}$

Copy and complete the magic square.

Tip

Copy the diagram and fill in the cells as you work them out.

$3 = \dfrac{a}{b} + c$

a, b and c are integers.

a Suggest values of a, b and c that will make the calculation correct.

b Suggest values of a, b and c that will make the calculation correct where $b > 1$.

c Suggest values of a, b and c that will make the calculation correct where $b < 0$.

There are approximately 5×10^9 red blood cells per millilitre of blood. James' blood volume is approximately 9 pints.

One pint is equal to 568 ml.

If James donates 10 per cent of his blood, how many red blood cells has he still got?

Alesha has a pack of playing cards. She picks one card at random, puts it on the table and then picks a second card.

Is it more likely that Alesha will get two aces, or two red kings?

How much more likely?

Tip

Would a diagram help? What about simplifying the problem, so she picks just one card?

Adam has two pairs of grey socks, three pairs of brown socks, a blue pair and a stripey pair. He also has four ties. His ties are pink, blue, purple and grey.

Adam takes a pair of socks and a tie at random each morning.

What is the probability that:

a he takes a pair of grey socks and the pink tie?

b he takes the stripey socks and the purple tie?

c he takes grey socks and a matching tie?

d he takes socks and a tie of the same colour?

e he takes socks and a tie of different colours?

> **Tip**
>
> What sort of diagram might be helpful here?

Rabbits can have from 4 to 8 kits (baby rabbits) in each litter.

The probability for the number of kits any female rabbit might have in a litter is given in the table.

Number of kits per litter	4	5	6	7	8
Probability	0.1	0.2	0.3	0.3	x

What is the probability that two randomly picked female rabbits have a total of ten kits in one breeding season?

Juliet is inviting some friends for afternoon tea. She wants to make chocolate brownies and lemonade.

In her mother's cupboard there is 350 g of sugar. Each glass of lemonade needs 30 g of sugar and each brownie requires 20 g of sugar.

Let x represent the number of glasses of lemonade and y represent the number of brownies.

a Write an inequality to show how much lemonade and how many brownies Juliet could make.

Juliet expects her friends will each have one glass of lemonade and at least one brownie.

b Write an inequality to show the number of friends Juliet could invite.

17

Harvey carried out a combination of transformations on a starting shape on a centimetre-square grid:

Translate by the vector $\begin{pmatrix} 2 \\ -1 \end{pmatrix}$.

Rotate 90° clockwise about (1, 0).

All but Harvey's final image is hidden under a piece of paper.

What will the diagram look like when the paper is lifted?

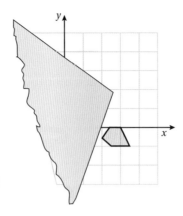

Tip

How can you work backwards through this problem? What are the 'inverse' transformations that need to be applied?

18

a Say how the pair of triangles shown below could be used to show that $\sqrt{8} = 2 \times \sqrt{2}$.

Diagrams not to scale

 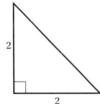

b Draw a pair of triangles to demonstrate that:

$\sqrt{18} = 3 \times \sqrt{2}$

Tip

What can you work out?

Tip

Why is it useful that the two triangles are right-angled? The two triangles are mathematically similar. What does this mean and how can it help you?

19

A field is 80 m long by 70 m wide. A goat is tethered to a pole in the centre of the field by a rope 15 m long.

A farmer plants wheat across the width of the field for 30 m from one end. The goat eats all the wheat it can reach.

It costs £65 to plant a 1 m strip of wheat the width of the field, and the farmer makes a return of £4.50 per square metre harvested. The farmer plants strips the complete width of the field but does not move the goat.

The farmer still makes a profit. How much profit?

A piece of paper is folded to make a kite as shown in the diagram.

The side lengths of the original piece of paper are in the ratio $1 : \sqrt{2}$.

Calculate the perimeter of the kite.

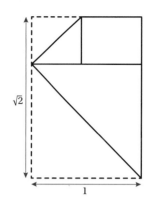

Diagram not to scale

The formula for the kinetic energy (in joules, J) of an object of mass m (in kg), travelling at a speed v (m/s), is:

$$KE = \frac{1}{2}mv^2$$

Two students calculate the kinetic energy of a tennis ball of mass 56 g after it is served at a speed of 120 miles per hour.

Their answers are shown below:

Student 1:

$$KE = \frac{1}{2}mv^2 = \frac{1}{2} = 56 \times 120^2 = 403\ 200\ J$$

Student 2:

Tip

Try answering the question yourself.

The mass of the tennis ball needs to be changed into kg.

56 g = 0.056 kg

The speed of the tennis ball needs to be changed into metres per second (m/s). Using the conversion 5 miles = 8 km:

$$\frac{120}{5} = 24$$

24 × 8 = 192 so 120 mph = 192 km/h
192 km/h = 192 000 m/h
There are 60 × 60 = 3600 seconds in one hour.

$$\frac{192\ 000}{3600} = 53.3333...\ m/s$$

KE = 0.5 mv²
0.5 × 0.056 × 53.3 = 1.4924
1.49242 = 2.227...
So the KE = 2.2 J.

Both students have done some things correctly in their calculations but have ended up with the wrong answer.

Say how each student should alter their method to achieve the correct result of 79.6 joules.

1

	Hot chocolate	Tea	Coffee	Total
Women	**8**	9	**12**	29
Men	8	7	**10**	**25**
Total	16	**16**	22	54

Of the 54 workers surveyed, 22 preferred coffee, 16 preferred tea and 16 preferred hot chocolate.

Peter was incorrect: tea is not the most popular drink. Coffee is the most popular hot drink among the workers, and hot chocolate is as popular as tea.

A diagram like this 2-way table will help here. Fill in the information you are given (in bold here), then work out the remaining figures.

2

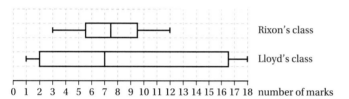

0 1 2 3 4 5 6 7 8 9 10 11 12 13 14 15 16 17 18 number of marks

A boxplot might be helpful here.

The highest score in Rixson's class is 12 and the highest score in Lloyd's class is 18. The best person in Lloyd's class did much better than the best person in Rixson's class.

The median for both classes is very similar. This tells you that half of Rixson's class got 7.5 or more and half of Lloyd's class got 7 or more marks.

The mean mark in Lloyd's class was bigger than the mean mark for Rixson's class, and this, combined with the bigger inter quartile range, tells you that there must have been several very high scores in Lloyd's class. In fact, the upper quartile for Lloyd's class must be better than the best score in Rixson's class, which means at least three of Lloyd's students did better than Rixson's best student.

At the bottom end it looks as if at least three of Lloyd's students did worse than Rixson's worst student.

It is difficult to tell, but because the means are different you can probably say that Lloyd's class did better.

3 **a**

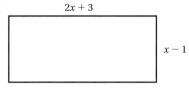

The perimeter = $2x + 3 + x - 1 + 2x + 3 + x - 1$

An expression for the perimeter is $6x + 4$.

b An expression for the area is $(2x + 3)(x - 1)$, which can be expanded and simplified to give $2x^2 + x - 3$.

c $2x^2 + x - 3 = 250$

$2x^2 + x - 253 = 0$
$(2x + 23)(x - 11) = 0$

$x = -11.5$, or $x = 11$

x cannot be negative (because that would mean at least one of the sides of the rectangle would be negative), so $x = 11$ and the two sides are $2x + 3 = 25\,$cm and $x - 1 = 10\,$cm.

The longest side is therefore 25 cm.

d The perimeter is $6x + 4$. Substituting $x = 11$ in this expression gives a perimeter of 70 cm.

To work out the perimeter of a rectangle, you need to add up the sides.

A diagram will help.

To work out the area you multiply the two sides.

Make an equation and solve it.

Factorise.

4 **a** 4 metres

b 8 bricks

c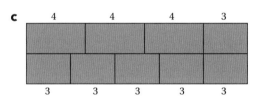

Option 1: $4\,$m $+ 4\,$m $+ 4\,$m $+ 3\,$m

Option 2: $3\,$m $+ 3\,$m $+ 3\,$m $+ 3\,$m$+ 3\,$m

d Option 1: 19 bricks

Option 2: 20 bricks

One less than five bricks = 4

Length + 1 = number of bricks

Any combination of 4 and 3 to total 15. Draw a diagram to help.

$4\,$m $+ 4\,$m $+ 4\,$m $+ 3\,$m would need: $5 + 5 + 5 + 4$ bricks (for every metre of plank there is one extra brick)

or

$3\,$m $+ 3\,$m $+ 3\,$m $+ 3\,$m $+ 3\,$m would need: $4 + 4 + 4 + 4 + 4$ bricks

5

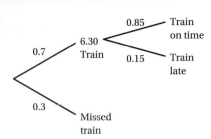

A tree diagram might be useful here.

Probability Leela is on time is $0.7 \times 0.85 = 0.595$

6 a

Here is a diagram that will work well.

+	1	2	3	4	5
1	2	3	4	5	6
2	3	4	5	6	7
3	4	5	6	7	8
4	5	6	7	8	9
5	6	7	8	9	10

b 10

c $\dfrac{3}{25}$

7 a

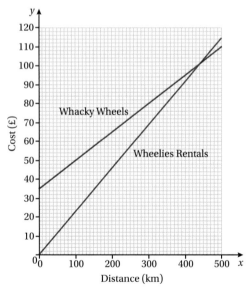

The graph is helpful in answering the rest of this question.

b Wheelies Rentals

c Yes, she would have saved £5.

Whacky Wheels: £35 + £75 = £110

Wheelie Rentals: 5 × £23 = £115

8

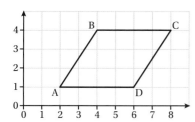

a $\overrightarrow{CD} = \begin{pmatrix} -2 \\ -3 \end{pmatrix}$

A diagram of the situation helps you see what is going on.

b $\overrightarrow{AB} = \begin{pmatrix} 2 \\ 3 \end{pmatrix}$

The length AB is the same as the length CD but the vectors are in opposite directions.

c ABCD is a parallelogram – a four-sided shape with two sets of parallel sides.

BC is $\begin{pmatrix} 4 \\ 0 \end{pmatrix}$ and AD is $\begin{pmatrix} 4 \\ 0 \end{pmatrix}$ so BC is parallel to AD.

AB is $\begin{pmatrix} 2 \\ 3 \end{pmatrix}$ and DC is $\begin{pmatrix} 2 \\ 3 \end{pmatrix}$ so AB is parallel to DC.

9 The diameter of the cherry tree is 3 m. I will assume it has a radius of 1.5 m. This means the tree can be planted anywhere 1.5 m away from the edge of the lawn.

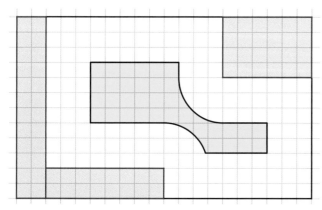

This diagram has a scale of two squares to 1 m. The yellow area is where the cherry tree can be planted.

You can choose any scale you like.

To show where the cherry tree can go there are straight lines 1.5 m away from the edge of the lawn and arcs (parts of circles) at the corners of the flower bed and the vegetable patch.

10 The helicopters cover 25 km from each town, so we need two circles. The fire brigade covers the section closer to B than A, so we need the locus of points equidistant from both towns.

> This diagram uses a scale of 1 cm to 5 km but you could choose a different scale. Start with a line of 8 cm to represent the 40 km distance between the towns.

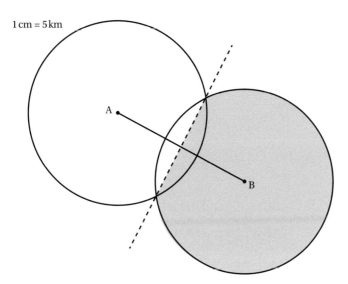

> With compasses draw a circle centre B and radius 5 cm.
>
> With compasses draw a circle centre A and radius 5 cm.
>
> The locus of points equidistant from B and A is the perpendicular bisector of line AB.
>
> Shade the region that is closer to B than A and covered by helicopter B.

11

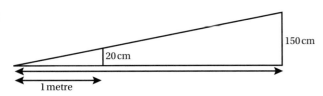

> A good diagram shows that this is a question about similar triangles.

The scale factor of the enlargement is
150 cm ÷ 20 cm = 7.5

1 m × 7.5 = 7.5 m

The image needs to be 7.5 larger to fill the screen. This means that the projector needs to be 7.5 times further away.

12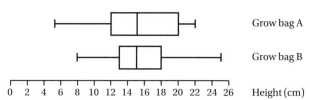

Boxplots are useful diagrams to use here.

a Grow bag A: the tomato plants vary quite a lot in height from 5 cm to 22 cm. The middle 50% shows a range of 8cm. The smallest plants are in bag A.

Grow bag B: the tomato plants are more consistent in size with a smaller range and much smaller IQR, although the biggest plant is in bag B.

b Because Elspeth is hoping to sell the tomato plants, she should use Grow bag B so most plants would be around the same size.

 13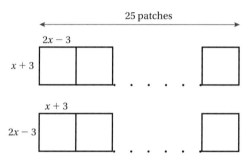

There are 25 patches in each row.

The width could be $25(x + 3) = 25x + 75$

or it could be $25(2x - 3) = 50x - 75$

a $25x + 75$ or $50x - 75$

b $64x - 96$ or $32x + 96$

$32(2x - 3)$ or $32(x + 3)$

c The two possible expressions for the area are $(25x + 75)(64x - 96)$ and $(50x - 75)(32x + 96)$.

These are equal because they are both equivalent to working out $25(x + 3) \times 32(2x - 3)$.

This is $800(2x^2 + 3x - 9)$, which is the same as $1600x^2 + 2400x - 7200$.

d $2.8\,m^2$ is the same as $2.8 \times 100\,cm \times 100\,cm = 28\,000\,cm^2$.

The equation is therefore: $1600x^2 + 2400x - 7200 = 28\,000$

Dividing through by 800 gives: $2x^2 + 3x - 9 = 35$

$2x^2 + 3x - 44 = 0$ can be factorised to give $(2x + 11)(x - 4) = 0$, so $x = -5.5$ (which isn't possible for this scenario) or $x = 4$.

The dimensions of the patches are: $2x - 3$ by $x + 3$ and when $x = 4$ this gives 5cm by 7cm.

14 a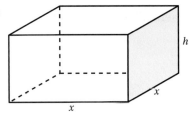

Draw a diagram of the container.

The volume is $x^2 \times h$. This equals 1440, so:

$x^2 h = 1440$, which means $h = \dfrac{1400}{x^2}$

b The base has area x^2. Each of the side walls (shown in yellow on the diagram) has area hx, which is $\dfrac{1400}{x^2} x$, which simplifies to give $\dfrac{1400}{x}$. There are four of these, so the total area of paper is $x^2 + \dfrac{5760}{x}$

c 12 cm by 12 cm

$1440 = x^2 \times 10$

$144 = x^2$

$x = 12$

d 624 cm³

15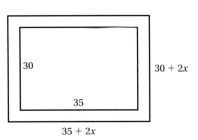

Draw a diagram and label it with what you know.

This is the top view of the pool and the path.

a $(35 + 2x)(30 + 2x) - 35 \times 30 = 130x + 4x^2$

You could work out the area of the outer rectangle and then subtract the area of the inner rectangle.

b Area of path = 3196.80 ÷ 30 = 106.56

Number of square metres of border

$4x^2 + 130x = 106.56$

Solve this equation.

$x = 0.8$

The path is 0.8 m wide.

16

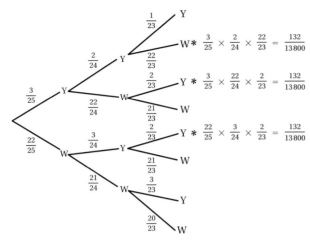

A probability tree diagram is probably most useful.

a $P(3 \text{ yellow}) = \dfrac{3}{25} \times \dfrac{2}{24} \times \dfrac{1}{23} = \dfrac{6}{13\,800} = \dfrac{1}{2300}$

b $P(3 \text{ white}) = \dfrac{22}{25} \times \dfrac{21}{24} \times \dfrac{20}{23} = \dfrac{9240}{13\,800} = \dfrac{77}{115}$

c The three routes labelled on the diagram all work, so you need to add them and will get:

$$\dfrac{132}{13\,800} + \dfrac{132}{13\,800} + \dfrac{132}{13\,800} = \dfrac{396}{13\,800} = \dfrac{33}{1150}$$

17

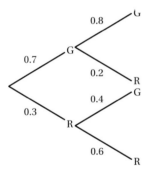

A tree diagram might help.

a $0.7 \times 0.8 = 0.56$

b $(0.7 \times 0.2) + (0.3 \times 0.4) = 0.26$

Two outcomes: green then red (GR) and red then green (RG).

18

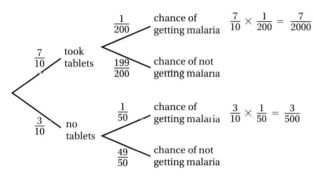

Badminton   Squash

G wins both: 0.85×0.35

G wins badminton,
R wins squash: 0.85×0.65

R wins badminton,
G wins squash: 0.15×0.35

R wins both: 0.15×0.65

> Draw the tree diagram.

a $0.85 \times 0.35 = 0.2975$

b $0.15 \times 0.35 = 0.0525$

c $0.85 \times 0.65 + 0.15 \times 0.35 = 0.605$

19

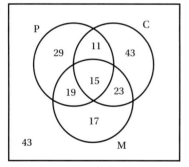

$\frac{1}{200}$ chance of getting malaria $\frac{7}{10} \times \frac{1}{200} = \frac{7}{2000}$

$\frac{7}{10}$ took tablets

$\frac{199}{200}$ chance of not getting malaria

$\frac{1}{50}$ chance of getting malaria $\frac{3}{10} \times \frac{1}{50} = \frac{3}{500}$

$\frac{3}{10}$ no tablets

$\frac{49}{50}$ chance of not getting malaria

> A tree diagram is helpful.

$$\frac{7}{2000} + \frac{3}{500} = \frac{19}{2000} = 0.0095$$

20 **a**

> A Venn diagram is likely to be useful here.
>
> Fill in the numbers you know, then work out the missing numbers.
>
> 53 study two subjects – so 11 study Physics and Chemistry but not Maths.
>
> 74 study Physics, so 15 study all three.
>
> 92 study Chemistry, so 43 study only Chemistry.
>
> There are 200 students so 17 study only Maths.

b $\frac{17}{200}$

c $\frac{68}{200} = \frac{17}{50}$

> 11 studied Chemistry and Physics, 19 studied Physics and Maths, 23 studied Chemistry and Maths, 15 studied all three.
>
> Total: 68 of 200 students.

21

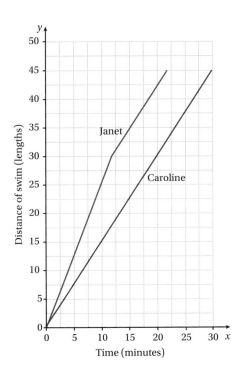

This is the sort of question where a graph can really help.

a 15

b 10 minutes

c $45 \div 30 = 1.5$ lengths per minute

d 8 minutes

e About 2 lengths per minute

f 15 more lengths, which is 60 in total, but she might get tired and start to slow down.

22 **a**

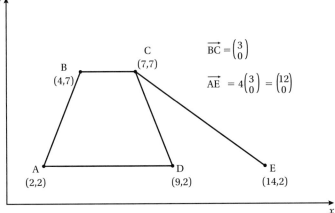

An isosceles trapezium has one line of symmetry, so D could be (9, 2) as C to D will be 2 units to the right and 5 units downwards.

b $\overrightarrow{AC} = \begin{pmatrix} 5 \\ 5 \end{pmatrix}$

c E (14, 2)

d $\overrightarrow{BE} = \begin{pmatrix} 10 \\ -5 \end{pmatrix}$

23

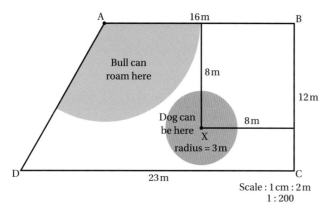

A good way to solve this is to draw a scale diagram. You could use a scale of 1 cm to 2 m.

On the scale diagram, construct:

* A circle of radius 1.5 cm about X to show where the dog can go.

* A locus of radius 4 cm about A between the fence lines DA and AB where the bull can roam.

Any path between D and B that does not enter either of the shaded areas would be safe to use.

24

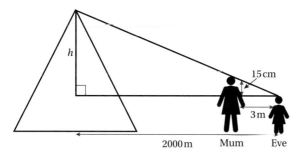

This is a complicated scenario. A diagram will help you see what is going on. Note that the diagram is not to scale: the people would be much smaller in relation to the height of the hill so the height of the large right-angled triangle will be roughly the height of the hill.

The ratio of height of object : distance is roughly the same for both mum and the hill.

$0.15 : 3 =$

$h : 2000$

$0.15 \div 3 \times 2000 = 100 \, \text{m}$

Divide by 3 and multiply by 2000.

The hill is about 100 m high.

25   **a**   width : length

40 : 55

8 : 11

$112 \div 8 = 14$

Length $= 11 \times 14 = 154 \, \text{cm}$

b $148.5 \div 11 = 13.5$

Width $= 13.5 \times 8 = 108 \, \text{cm}$

c Width of photo = area ÷ length

= 127.5 ÷ 15

= 8.5 cm

Factor of enlargement

= new width ÷ original width

= 25.5 ÷ 8.5

= 3

The area will increase by a factor of $3^2 = 9$.

Area of poster = $(127.5 \times 9)\,\text{cm}^2$

$= 1147.5\,\text{cm}^2$

First work out the width of the original photo using the length and area given in the question.

Then work out the scale factor of the enlargement by comparing the two widths.

26 **a i, ii**

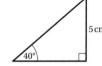

There are three possible triangles that fit Harriet's description.

iii She could say that the hypotenuse is 5 cm.

For only one triangle to be possible, Harriet must make sure that her conditions follow one of the conditions of congruence:

SSS, SAS, ASA, AAS or RHS

b i

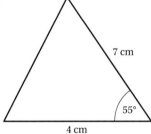

There is only one triangle that satisfies Janet's conditions because she has given SAS.

ii A scale drawing will enable you to measure the length of the third side 5.7 cm to the nearest mm.

$a^2 = b^2 + c^2 = 2bc \cos A$

$a^2 = 7^2 + 4^2 - 2 \times 7 \times 4 \times \cos 55$

$a = 5.734...$

The third side is 5.7 cm.

A more accurate method to find the length of the third side is to use the cosine rule.

27 **a**

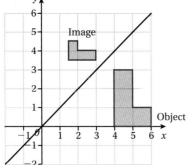

Choose a shape without any symmetry to easily see what is going on.

b

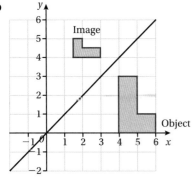

The orientation and size of the final image are the same when the transformations are carried out in a different order, but it ends up in a different place.

c 1, 2, 3

 1, 3, 2

 2, 1, 3

 2, 3, 1

 3, 1, 2

 3, 2, 1

Six different images.

Consider all the possible arrangements of the three transformations.

You have already drawn the first and last one.

The others result in an image that is the same size and same way around, but which is in a different place.

28 **a**

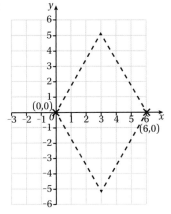

$a^2 + b^2 = c^2$

$3^2 + b^2 = 6^2$

$9 + b^2 = 36$

$b^2 = 27$

$b = \sqrt{27} \, (= 5.196...)$

$(3, \sqrt{27}), (3, -\sqrt{27})$

b

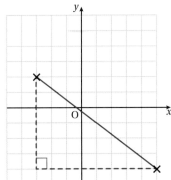

Here is a diagram showing the two points.

This time you only need to know the length of the sides. Pythagoras tells you the side has a length of 10.

$6^2 + 8^2 = c^2$

$c^2 = 100$

$c = \sqrt{100} = 10$

Now, if you have an equilateral triangle of side 10:

$5^2 + b^2 = 10^2$

$25 + b^2 = 100$

$b^2 = 75$

$b = \sqrt{75} \, (= 8.660...)$

Area $= \dfrac{1}{2} \times 10 \times \sqrt{75}$

Area $= 43.301...$

43.3 square units

There are two possible locations for the third vertex.

The x-coordinate is 3.

Use Pythagoras' theorem to calculate the vertical height of the equilateral triangle.

Calculate the area of the equilateral triangle ($\dfrac{1}{2} \times$ base $\times$ height).

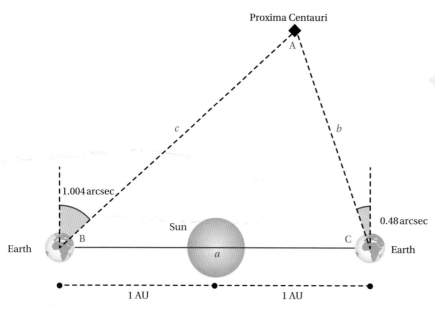

There are lots of aspects of this question that aren't particularly clear, so you will need to make some assumptions (which is fine as long as you explain them) and a good diagram will be a very important starting point.

It looks as if the angles are measured perpendicularly from the line through the Sun, so
angle B is 90° – 1.004 arcsec,
angle C = 90° – 0.48 arcsec, and
angle A is 1.004 + 0.48 = 1.484 arcsec

It is not obvious which distance is being asked for – so let's work out lengths b and c.

$$\frac{a}{\sin A} = \frac{b}{\sin B} = \frac{c}{\sin C}$$

$$b = \frac{2 \text{AU}}{\sin(1.484 \text{ arcsec})} \times \sin(90° - 1.004 \text{ arcsec})$$

This gives $b = 4.169773711 \times 10^{13}$ km

$$c = \frac{2 \text{AU}}{\sin(1.484 \text{ arcsec})} \times \sin(90° - 0.24 \text{ arcsec})$$

This gives $c = 4.169773711 \times 10^{13}$ km

The distance is therefore about 4.2×10^{13} km

These answers are the same for the first 10 significant figures (which makes sense because Proxima Centauri is very, very far away compared to the diameter of the orbit of the Earth).

1 You know that $\sqrt{64} = 8$ and $\sqrt{81} = 9$, so $8 < \sqrt{70} < 9$

> You don't know how big $\sqrt{70}$ is, so think about the closest square roots that you do know.

Looking at the statement in the question:

$10\sqrt{70} > 8 \times 10$ so the statement is incorrect.

---

2 $5 \times 12 = 60$

> If the mean of five numbers is 12, what do you know? When you work out the mean of five numbers you add them and divide by 5, so before the sum was divided by 5 it must have been 60 (because $60 \div 5 = 12$).

$x + x + 3x + 4x + 6x = 60$

> Use the ratios to form an equation.

$$15x = 60$$
$$x = \frac{60}{15} = 4$$

The five numbers are 4, 4, 12, 16 and 24, so the largest number is 24.

3

45°

> Because of the way the paper has been folded you know the marked angle is 45°.

$360° \div 45° = 8$, so 8 of the kites will fit together.

---

4 $835 \div 16.5 = 50.60$

> If the figures were exact then you would work out $840 \div 16$. Here, to be safe, you need to assume the van can only cope with 835 kg and that the slabs weigh 16.5 kg.

He can transport 50 slabs.

5 a

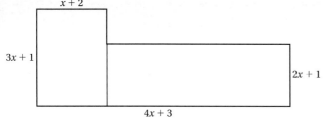

Area $= (3x + 1)(x + 2) + (3x + 1)(2x + 1)$

$\qquad = (3x + 1)(3x + 3)$

$\qquad = 9x^2 + 12x + 3$

First, divide the area into two rectangles.

Work out the area of each rectangle and add them to find the total area.

$4x + 3 - (x + 2) = 3x + 1$

b $3x + 1 = 16\,\text{m}$, so $3x + 3 = 18\,\text{m}$

Area $= 16 \times 18 = 288\,\text{m}^2$

Use the factorised form of the area.

c 78 posts

Work out the perimeter, $14x + 8$, with $x = 5$.

d $78 \times 3 = 234\,\text{m}$ of wire

e $(78 \times 18.50) + (234 \times 2.30) = £1981.20$

6

You might not be able to answer this immediately, but you can make some simultaneous equations and solve them.

6 mangoes and 4 pawpaws cost £5.80.

Use the first sentence and double the amounts.

5 mangoes and 4 pawpaws cost £5.50, so one mango must cost 30p.

Use the second sentence.

The price of a pawpaw is $\dfrac{(£5.50 - £1.50)}{4} = £1$.

Substitute the value of 30p for a mango into one of the earlier equations to work out the cost of a pawpaw.

---

7 a p(green ball) $= 0.64$, so p(not a green ball) $= 0.36$

You can't solve this immediately, but you do know the probability of not picking a green ball.

The probability of getting a white ball is the same as that of getting a yellow ball, so they must each be $0.36 \div 2 = 0.18$.

b 9 yellow balls out of the total number $= 0.18$

$0.18 = \dfrac{18}{100} = \dfrac{9}{50}$ so there are 50 balls in total.

---

8 The probability that Sharon or Frances finishes first is
0.23 + 0.15 = 0.38

So the probability that Sue or Anisha finishes first is
1 − 0.38 = 0.62

You don't need to work out the probability that Anisha finishes first.

9

Child	Number of sixes in 10 rolls	Relative frequency
A	7. Total number of sixes is 7.	0.7
B	4. Total number of sixes is 11 out of 20.	0.55
C	3. Total number of sixes is 14 out of 30.	0.47
D	6. Total number of sixes is 20.	0.5
E	4. Total number of sixes is 24.	0.48

It probably makes sense to start working out the probability after each child has had a go.

Now you might want to skip a few. After 10 children have done their throws, there are 44 sixes and the relative frequency is 0.44.

After 15 children have done their throws, there are 68 sixes and the relative frequency is 0.45.

After 20 children have done their throws, there are 83 sixes and the relative frequency is 0.415.

The relative frequency is going up and down, but it looks as if it is settling somewhere close to 0.4.

When more trials happen the relative frequency is generally closer to the actual value.

The actual value of the probability is probably about 0.4, which means that the dice is a biased one.

10 **a**

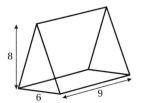

First, label a copy of the diagram.

The area of the triangular cross-section is
$\frac{1}{2} \times 6 \times 8 = 24\,\text{cm}^2$

The volume of the prism is $24 \times 9 = 216\,\text{cm}^3$

b The volume of the pentagonal prism is
$36 \times L = 216\,\text{cm}^3$

$L = 216 \div 36 = 6\,\text{cm}$

11

Volume of cylinder $= \pi \times 3^2 \times 11 = 311.01767... \, cm^3$

$\qquad\qquad\qquad\qquad\quad = 311 \, cm^3 \, (3 \, sf)$

$311 \, cm^3 = 311 \, ml$

Volume of water left in bottle $= 500 \, ml - 311 \, ml = 189 \, ml \, (3 \, sf)$

> You could start by working out the volume of a cylinder of water that is 11 cm high.

> Volume of water (a cylinder) = area of circular base × height

> Write the volume of water in ml.

> Volume of water left in bottle

 12　**a**　$3:5$ gives 8 parts.

$32 \div 8 = 4$

There are $3 \times 4 = 12$ non-calculator questions on each test and $5 \times 4 = 20$ calculator questions.

b　15 tests $\times$ 20 = 300 calculator questions

c　$15 \times 32 = 480$ questions, so each teacher will write 240 questions.

Miss Smith will write $15 \times 12 = 180$ non-calculator questions, so she will have to write $240 - 180 = 60$ calculator questions.

$\dfrac{60}{240} = \dfrac{1}{4}$

> Start by working out the number of parts in the ratio.

> This is complicated, so let's find out how many questions there are altogether.

 13　**a**　$40 \times 30 = 1200$ unit squares

$\dfrac{4}{5}$ of $1200 = 960$

You want to have $\dfrac{5}{6}$ of $1200 = 1000$ shaded, so you need to shade an extra 40 unit squares.

b　960 is shaded from the first rectangle.

$\dfrac{2}{3}$ of $1200 = 800$ is shaded from the second rectangle.

The fraction of the whole thing that is shaded is:

$\dfrac{960 + 800}{1200 + 1200} = \dfrac{1760}{2400} = \dfrac{11}{15}$

c　$\dfrac{3}{4}$ of $2400 = 1800$, so an extra 40 unit squares need to be shaded.

> You could start by working out the area of the rectangle.

> Now you can work out the shaded area.

> One way to do this is to work out the total area and the total amount that is shaded.

> You already know that 1760 unit squares are shaded.

14 $6 \div 0.5 = 12$

$6 \div 0 =$ Error

There is a link between $20 \div 4 = 5$ and $4 \times 5 = 20$.

$6 \div 0.5 = x$ means that $0.5 \times x = 6$.

This makes it clear that $x = 12$.

$6 \div 0 = y$ means that $0 \times y = 6$.

It is not possible to multiply 0 by something and to get the answer of 6, so $6 \div 0$ does not have an answer.

> There are lots of ways to explain why. Here is one.

---

15   **a** The angles in a regular hexagon are 120°, and those in a regular pentagon are 108°.

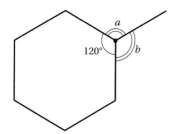

> In a regular shape all the angles must be the same so working these out might be a good starting point.

The angles around a point should add up to 360°, but here they total $120° + 216° = 336°$, pentagons cannot be used to fit round the hexagon.

b $3 \times 120° = 360°$, so regular hexagons will fit around the edge. Equilateral triangles would also fit:

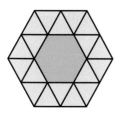

c Two regular pentagons have angles of
$108° + 108° = 216°$.

This leaves $360° - 216° = 144°$,

which is the interior angle of a regular decagon.

A regular decagon can be surrounded by regular pentagons.

> The interior angle of a regular pentagon is 108°.
>
>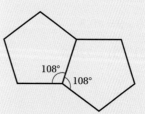
>
> You have to identify the regular polygon that has an interior angle of 144°.
>
> The exterior angle is $180 - 144 = 36$
>
> Exterior angles always sum to 360°, so there must be 10 of them and the shape must have 10 sides.

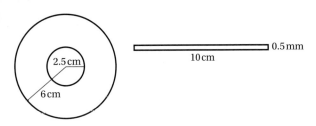

The diagrams show the end of the roll of paper and the side view of a single piece of paper.

The area of the 'ring' = $\pi \times 6^2 - \pi \times 2.5^2 = 93.46238\,\text{cm}^2$

The area of the side of one piece of paper = $10 \times 0.05 = 0.5\,\text{cm}^2$

$93.46238 \div 0.5 = 186.92476$

This means there are approximately 187 sheets on the roll. You cannot tell whether this is exact because the sheets might be compressed when they are rolled up.

If there are 200 sheets then they must be slightly thinner, so they are not as high quality as was claimed.

The width of the paper is irrelevant. You only need to worry about the area of the end of the roll.

Convert 0.5 mm into 0.05 cm.

Work out how many pieces of paper have the same area as the ring.

  a

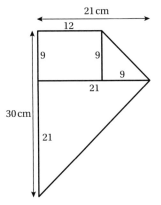

Rectangle: $12 \times 9 = 108$

Small triangle: $\frac{1}{2} \times 9 \times 9 = 40.5$

Large triangle: $\frac{1}{2} \times 21 \times 21 = 220.5$

Total area = $369\,\text{cm}^2$

b Area of the original sheet is $21 \times 30 = 630\,\text{cm}^2$

The fraction is $\dfrac{369}{630} = \dfrac{41}{70}$

Draw a good diagram and then work out the lengths of the sides.

The two right-angled triangles are each half a square and this helps you to work out the lengths.

Now there is enough information to work out the areas.

Either work out the area of the rectangle and the two triangles and add them, or work out the area of the original sheet of paper and subtract two triangles.

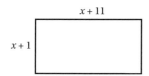

You can't immediately work out the size (using numbers) of the sides of the rectangle, but you know that the area of the rectangle is the same as the area of the square, so start by working those out.

The area of the rectangle is
$(x + 1)(x + 11) = x^2 + 12x + 11$

The area of the square is $(x + 5)^2 = x^2 + 10x + 25$.

These are equal, so $x^2 + 12x + 11 = x^2 + 10x + 25$.

This simplifies to give $2x = 14$, so $x = 7$.

The dimensions of the rectangle are therefore 8 cm by 18 cm.

$18 \times 8 = 144$ cm² so 144 centimetre squares will be needed to cover the rectangle.

 a The big square has area $5x \times 5x = 25x^2$

Subtracting the area of the small square (x^2) gives $24x^2$ as the expression for the blue area.

b $24x^2 = 1944$, so $x^2 = 81$, which means that here $x = 9$

You can't immediately see what the perimeter will be, so start by working out the value of x.

The perimeter of the shape is $20x$, which equals 180 cm.

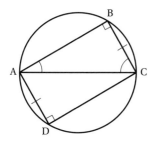

$A\hat{B}C = 90°$ and $A\hat{D}C = 90°$

Because the angle in a semicircle is a right angle.

$AD = BC$

Compare triangles ABC and CDA.

AC is common to both triangles and is the hypotenuse.

So, by RHS, triangles ABC and CDA are congruent.

This means $B\hat{A}C = A\hat{C}D$. But $B\hat{A}C + A\hat{C}B = 90°$ (because angles in a triangle add up to 180°), so $D\hat{C}B = 90°$

For a similar reason, $D\hat{A}B = 90°$

All four angles in the quadrilateral ABCD are 90°, so it is a rectangle.

21 **a**

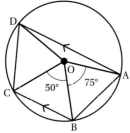

$\hat{OCB} = \hat{OBC} = (180° - 50°) \div 2 = 65°$

Each of the four triangles is isosceles (because the lines OA, OB, OC are radii).

$\hat{OAB} = \hat{OBA} = (180° - 75°) \div 2 = 52.5°$

Use this to calculate the other angles in triangles BOC and AOB.

$\hat{CBA} = 65° + 52.5° = 117.5°$

$\hat{BAD} = 180° - 117.5° = 62.5°$

Because AD is parallel to CB (trapezium), DAB + ABC = 180°.

$\hat{OAD} = \hat{ODA} = 62.5° - 52.5° = 10°$

Now calculate OÂD in the isosceles triangle OAD.

$\hat{AOD} = 180° - 20° = 160°$

Now calculate AÔD, remembering to subtract both OÂD and ODA.

$\hat{COD} = 360° - 50° - 75° - 160° = 75°$

Angles around a point sum to 360°.

b OA = OB = OC = OD because they are all a radius of the circle, and so, by SAS, triangle COD is congruent to triangle AOB. This means CD is equal to AB, so ABCD is an isosceles trapezium.

You need to show which angles you are looking at, so first draw in lines from each vertex of the trapezium to the centre of the circle and label the angles given in the question.

22

θ	$\sin\theta$	$\cos\theta$	$\sin\theta \div \cos\theta$
0°	0	1	0
30°	$\frac{1}{2}$	$\frac{\sqrt{3}}{2}$	$\frac{1}{2} \div \frac{\sqrt{3}}{2} = \frac{1}{2} \times \frac{2}{\sqrt{3}} = \frac{1}{\sqrt{3}}$
45°	$\frac{\sqrt{2}}{2}$	$\frac{\sqrt{2}}{2}$	1
60°	$\frac{\sqrt{3}}{2}$	$\frac{1}{2}$	$\frac{\sqrt{3}}{2} \div \frac{1}{2} = \frac{\sqrt{3}}{2} \times \frac{2}{1} = \sqrt{3}$
90°	1	0	$1 \div 0$, which is undefined

It seems sensible to start by following the instructions and filling in the missing boxes in the table!

The values in the final column do look like tan θ.

In a right-angled triangle: $\sin\theta = \dfrac{\text{opp}}{\text{hyp}}$ and $\cos\theta = \dfrac{\text{adj}}{\text{hyp}}$

This means that

$\dfrac{\sin\theta}{\cos\theta} = \dfrac{\text{opp}}{\text{hyp}} \div \dfrac{\text{adj}}{\text{hyp}} = \dfrac{\text{opp}}{\text{hyp}} \times \dfrac{\text{hyp}}{\text{adj}} = \dfrac{\text{opp}}{\text{adj}}$

$\dfrac{\text{opp}}{\text{adj}} = \tan\theta$,

so it is the case that $\sin\theta \div \cos\theta = \tan\theta$

$\frac{2}{16}$ had a dot, which is $\frac{1}{8}$.

That means the 16 he caught were about $\frac{1}{8}$ of the population, so there are $16 \times 8 = 128$ altogether, which is below the limit of 150.

As long as the woodlice he marked had mixed themselves up with the others, the fraction with dots the second time is the fraction he caught.

 In the original pentagon all the angles were 108° (angles in a regular pentagon) and in the equilateral triangles the angles are all 60°.

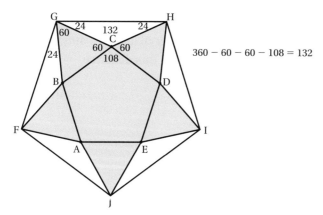

There are lots of ways to say how this works.

Draw the pentagon in and then look at the diagram. To show that the outer pentagon is regular you need to show that the angles are all equal and that the sides are all the same length.

Start with the angles.

The angles in the isosceles triangles around the edge are therefore 132°, 24°, 24°.

The angles in the outer pentagon are therefore 24° + 60° + 24° = 108°, which are the angles for a regular pentagon.

All the sides of the original pentagon are equal (because it is regular), so the equilateral triangles are all congruent. The isosceles triangles are all congruent (by SAS) so the sides of the outer pentagon are all the same.

Because the sides are all equal and the angles are equal the outer pentagon is regular.

1 minute = 60 s

First calculate the number of seconds in 1 year.

1 hour = 60 minutes = 60 × 60 s = 3600 s

1 day = 24 hours = 24 × 3600 s = 86 400 s

1 year = 365 days = 365 × 86 400 s = 31 536 000 s

2.99×10^5 km/s × 31 536 000 s = 9.429... × 10^{12} km

Now calculate the number of km travelled by light in one year.

$4.22 \times 9.429... \times 10^{12}$ km = $3.979... \times 10^{13}$ km

Convert 4.22 light years into km.

$3.979... \times 10^{13}$ km = $3.979... \times 10^{16}$ m

Convert this into metres.

$3.979... \times 10^{16}$ m ÷ 400 m = $9.947... \times 10^{13}$

Calculate the number of laps of a 400 m track that are required to cover this distance.

You would need to run approximately 10×10^{13} or 1×10^{14} or 100 000 000 000 000 laps of a standard 400 m running track to cover the distance from Earth to Proxima Centauri.

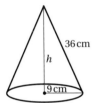

Start with a labelled diagram of the hat.

Circumference of the base is $\pi \times 2r = 18\pi$, and when the hat is unrolled, this is the length of the arc.

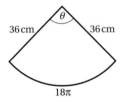

This is a sector of a circle.

The radius of the sector = 36 cm, so the circumference of a full circle would be $72\,\pi$.

The arc is $18\,\pi$, which is $\frac{1}{4}$ of the circumference, so the angle is $\frac{1}{4}$ of 360° = 90°.

Hence the sector required will be a quarter of a circle of radius 36 cm.

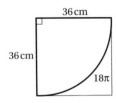

This will fit on a square piece of card of side length 36 cm.

27 Radius of large bowl = 16 cm ÷ 2 = 8 cm

Volume of large bowl = $\frac{4}{3} \times \pi \times 8^3 \div 2$ = 1072.93 cm³ (2 dp)

Volume of bowl (hemisphere) = $\frac{4}{3}\pi r^3 \div 2$

Radius of small bowl = 11 cm ÷ 2 = 5.5 cm

Volume of small bowl = $\frac{4}{3} \times \pi \times 5.5$ cm³ ÷ 2
                      = 348.45 cm³ (2 dp)

1072.33 cm³ – 348.45 cm³ = 723.88 cm³ (2 dp)

Difference between volumes of larger and smaller bowls

723.88 ÷ 100 = 7.2388

Number of kcal in 723.88 of soup = 7.2388 × 59 = 427 kcal saved (to the nearest kcal)

 28

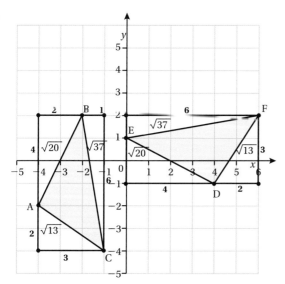

Draw an accurate version of the diagram on a grid.

Then work out the length of each of the sides, using Pythagoras' theorem.

The two triangles have identical sides, so by SSS they are congruent.

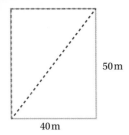

50 m

40 m

It seems sensible to work out the distances the two animals travel.

The tortoise travels $\sqrt{40^2 + 50^2} = \sqrt{4100}$

The hare travels $40 + 50 = 90$

You know how fast the tortoise moves, so you can work out how long it takes the tortoise to complete the race.

$$\text{Time} = \frac{\sqrt{4100}}{0.1} = 640.312 \text{ s}$$

Time = distance ÷ speed

The hare only needs to travel 90 m in 640.312 seconds, so the lowest speed it can go at and still beat the tortoise is $\frac{90}{640.312} = 0.1405 \text{ m/s}$

1

First you could work out the actual answer. Redraw the diagram with extra lines on it.

$\frac{9}{16}$ of the diagram is purple.

Team A have given an area rather than a proportion: 0 marks.

Team B have got the wrong fraction. Maybe they saw that $\frac{3}{4}$ of this part was shaded:

But that is not the whole thing: 0 or 1 mark

Team C are correct. Maybe they used the idea that the inner big white triangle is not included (leaving $\frac{3}{4}$ of the whole diagram) and then $\frac{3}{4}$ of that is shaded, which is $\frac{3}{4} \times \frac{3}{4}$. This equals $\frac{9}{16}$. 4 marks

Team D are correct. $\frac{9}{16} = 0.5625$: 4 marks

---

2 Approach 1:

The interior angle of an octagon = 135°

135° ÷ 2 = 67.5°

There are several ways to approach this problem.

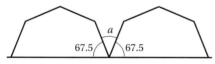

67.5° + 67.5° + a = 180°

a = 180° − 135° = 45°

Approach 2:

Making a copy of the half-octagons and rotating it creates two octagons. The angle a is the exterior angle.

The octagon has been cut in half so the angles adjacent to the one labelled a will each be half of the interior angle of the octagon.

The sum of angles on a straight line is 180°.

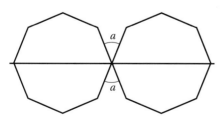

Exterior angle of an octagon = 360° ÷ 8 = 45°

3 **a**

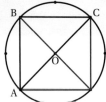

The inner shape is a square, so the angle ABC is 90°

Here are some extra lines that make it easier to see what is going on.

b

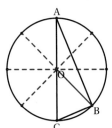

The extra lines show that there are 8 equal angles around the centre of the circle.

Each angle is therefore 360° ÷ 8 = 45°.

Triangle OCB is isosceles, which means angles B and C are equal and add up to 135°. Each one is therefore 67.5°.

Triangle OAB is also isosceles and the angle AOB = 135° (three lots of 45°). Angles A and B are equal and add up to 45°, so each one is 22.5°.

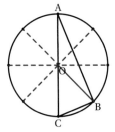

Angle ABC is 67.5° + 22.5° = 90°

c

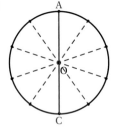

The angles around the centre are all 36° (360° ÷ 10). When you choose a point for B you will make isosceles triangles and can work out the other angles as in part **b**.

4 **a**

The extra lines on the diagram show that you need 180° plus a third of 180° (60°), which is 240° in total.

b When the minute hand makes a full turn the hour hand moves one hour, which means the hour hand moves at $\frac{1}{12}$ of the speed.

$\frac{1}{12}$ of 240° = 20°

It's a good plan to draw the angle you need to work out. Then a couple of extra lines will make the calculations easier.

5 **a**

The length of the fence is:

π × 17 = 53.407075 ... m

53.4 m

A copy of the diagram with the extra line added will help.

Circumference of circle = π × diameter

Round the answer to 3 sf.

b
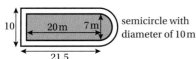
semicircle with diameter of 10 m

Length of fence:

21.5 + 10 + 21.5 + semicircle with diameter of 10 m

Semicircle:

π × 10 ÷ 2 = 15.70796 ... m = 15.7 m (3 sf)

Total length is 53 m + 15.7 m = 68.7 m

68.7 m – 53.4 m = 15.3 m

Calculate the difference in lengths of fencing.

Ian will need to buy 15.3 m more fencing than Gerry.

Conclusion.

141

6

650 m
130 m
a
a
130 m
650 m

Add some lines to make the shape into a rectangle.

The perimeter of the shape (which is the same as the perimeter of the rectangle) is

$a + a + 650 + 650 = 2a + 1300$

The perimeter is 2.28 km, so (after converting that to 2280 m):

$2a + 1300 = 2280$

Solving this gives $a = 490$ m, so the value of a is 0.49 km.

---

7 The sides of the triangle are all tangents to the circle.

A good extra line to try is one from the centre of the circle to the tangent point (because this does some special things, like forming right angles).

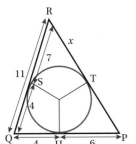

SQ = UQ, so SQ = 4.

RS is equal to RT, so $x = 7$.

You know this because two tangents from the same point are the same length.

8

a

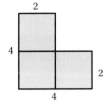

You need to divide the area into congruent shapes. Congruent shapes have the same area, so you could start by working out the area of the original diagram, which is 12 square units.

To have three congruent shapes the area of each one must be 4 square units. An easy way to make an area of 4 is to have a square.

b

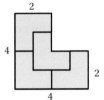

To make four congruent shapes the area of each one must be 3 square units. If you try to use three straight lines it doesn't work, either with rectangles or with triangles. There are other shapes that have an area of 3 square units, and these will work here.

9 **a** $3.6 \text{ m} = 360 \text{ cm}$

$360 \text{ cm} \div 18 \text{ cm} = 20$ stairs in the staircase.

The stairs must cover a vertical distance of 3.6 m. Each step is 18 cm high.

b Base length of staircase $= 20 \times 28 \text{ cm} = 560 \text{ cm}$

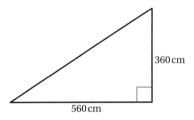

The banister rail will go diagonally up the stairs. It will need to be the same length as the diagonal length of the staircase. Use Pythagoras' theorem to calculate this.

$a^2 + b^2 = c^2$

$560^2 + 360^2 = c^2$

$443\,200 = c^2$

$c = 665.73... \text{ cm}$

The banister rail will need to be 6.66 m in length (to the nearest cm).

10 **a** The length is $6 - x$.

The perimeter is 12, so the length and the width together add up to 6.

The area is therefore $(6 - x)x$, which can also be written as $6x - x^2$.

b The area of the tabletop is equal to $6.3\,\text{m}^2$, so (using the area from part **a**):

The area of the tablecloth is big enough, but it might not be the right dimensions to fit neatly on the table, so you need to work out the dimensions of the table.

$6x - x^2 = 6.3$

$10x^2 - 60x + 63 = 0$

Rearrange and multiply through by 10.

$x = 4.64$ or $x = 1.36$

Use the quadratic formula to solve the equation.

These are the dimensions of the table (because $6 - 4.64 = 1.36$).

The tablecloth won't fit.

11

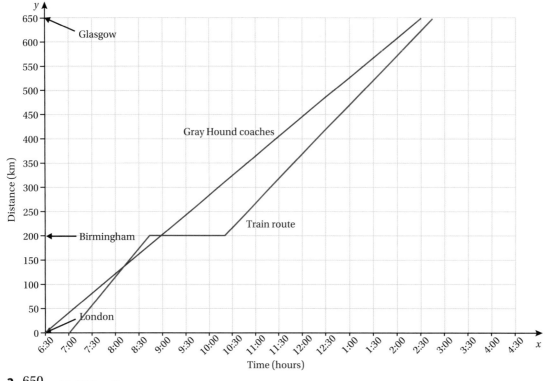

Even though you are not told to, it is helpful to draw the graph. Note that the graphs show 'average speeds' as the speeds will vary during the journeys.

a $\dfrac{650}{8} = 81.25 \text{ km/h}$

b 1 h 40 min + 1 h 45 min + 4 h 15 min = 7 h 40 min
$= 7\frac{2}{3}\text{ h}$

Work out the total time for the train journey in hours.

$650 \div 7\frac{2}{3} = 84.8 \text{ km/h}$

12 If Z, S and P are collinear then $\overrightarrow{SP}$ is parallel to $\overrightarrow{ZS}$ and they are multiples of each other.

$\overrightarrow{SP} = 2\mathbf{u} + 3\mathbf{v}$

$\overrightarrow{ZS} = \frac{1}{2}\mathbf{u} + 2\mathbf{v}$

$4\overrightarrow{ZS} = 2\mathbf{u} + 8\mathbf{v}$

There is no way to express $\overrightarrow{SZ}$ as a multiple of $\overrightarrow{SP}$.

Hence Z, S and P are not collinear.

It is worth adding these extra lines to the diagram.

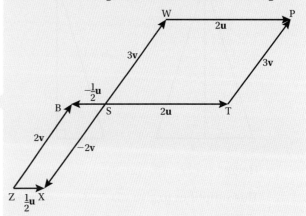

13 a

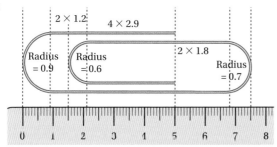

The three semicircular parts add up to:

$0.9\pi + 0.6\pi + 0.7\pi = 2.2\pi$

The straight parts add up to 17.6 cm.

The total length is 24.5 cm (1 dp).

Chop the diagram up into straight and semi-circular sections of wire and label them.

Work out the perimeter of a semicircle.

$\frac{2\pi r}{2} = \pi r$

b

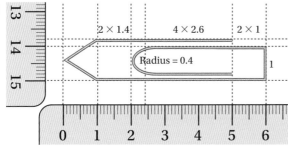

The diagonal parts are the hypotenuse of a right-angled triangle.

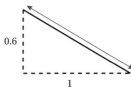

$h = \sqrt{1^2 + 0.6^2}$

The total length of wire is 19.8 cm (1 dp), which means the first paperclip contains more wire.

Use the same process as in part **a** to work out the length of the straight parts and the semicircle.

The total length will be: straight parts + diagonal parts + semicircle

145

14 **a**

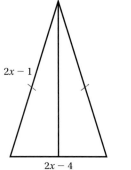

Put in the extra line – this makes two right-angled triangles.

Height $= \sqrt{(2x-1)^2 - (x-2)^2}$

This simplifies to give $\sqrt{3x^2 - 3}$

Work out the height of the triangle, using Pythagoras' theorem.

b $\sqrt{3x^2 - 3} = 12$

You are told the height is 12.

$3x^2 - 3 = 144$, so $x = 7$

The base of the original triangle is 10, so the area is 60 cm².

15 **a**

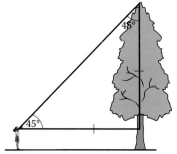

If they are looking up at an angle of 45°, then they create a right-angled, isosceles triangle. (The triangle is isosceles because if one angle is 90° and another is 45° then the remaining angle will also be 45°.)

Draw the extra lines on the diagram so that you can see what is going on.

Assuming that the child's height is small compared to the height of the tree, then the horizontal distance along the ground is approximately the same as the height of the tree (two equal sides in an isosceles triangle).

b

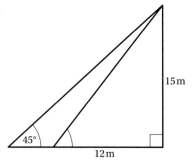

A good diagram will help.

Opposite side = height of tree = 15 m

Adjacent side = 12 m

Guy's angle $= \tan^{-1} \dfrac{15}{12} = 51.34 \ldots °$

Guy's angle is approximately 51°.

16

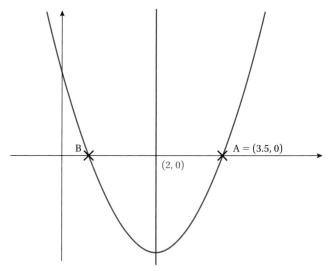

The question says that the graph is symmetrical about the line $x = 2$, so it makes sense to draw that line on a copy of the graph.

The coordinates of point B are (0.5, 0).

Now you can see that the crossing value for point A is 1.5 to the right of the line of reflection, so point B must cross at 1.5 to the left of the line.

C is (3.5, 0) and D is (0.5, 0).

In the second diagram the quadratic has been reflected in the x-axis so the crossing points are the same.

17

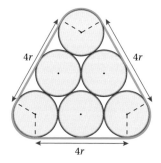

The diagram, with its extra lines, shows that 3 straight lines (each $4r$ in length) and 3 arcs are needed.

The 3 arcs together make a circle, so the total length is $12 \times r + 2\pi r$.

The radius is 2.7 cm, so the length is 49.36 cm.

147

18

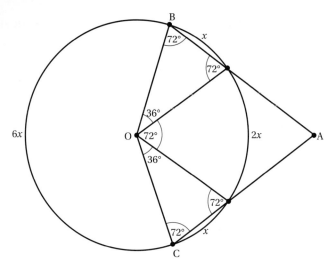

Draw in the radius of the circle. This gives two isosceles triangles.

The total circumference of the circle is $10x$, which means that x is $\frac{1}{10}$ of the circumference and the angle at the centre of the circle formed by each of the arcs of x is $360° \div 10 = 36°$.

The angle at the centre of the arc of $2x$ is therefore $72°$.

In quadrilateral ABOC the angles must add up to $360°$ so angle BAC must be $72°$.

Use the fact that these are isosceles triangles to work out the other angles in those triangles.

19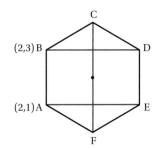

a $\overrightarrow{DE} = \begin{pmatrix} 0 \\ -2 \end{pmatrix}$

$\overrightarrow{DE}$ is the same vector as $\overrightarrow{BA}$. This is the same numbers as $\overrightarrow{AB}$ but with the opposite sign.

b $(2 + \sqrt{3}, 4)$

In a regular hexagon the interior angles are each 120°. By drawing a right-angled triangle where the hypotenuse side is of length 2 units, you can use the edge of the hexagon and trigonometry to calculate the other lengths.

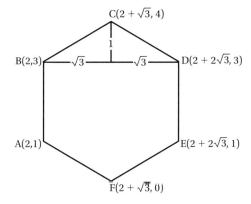

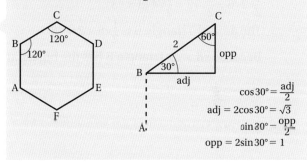

$\cos 30° = \dfrac{\text{adj}}{2}$

$\text{adj} = 2\cos 30° = \sqrt{3}$

$\sin 30° = \dfrac{\text{opp}}{2}$

$\text{opp} = 2\sin 30° = 1$

c $\begin{pmatrix} \sqrt{3} \\ 1 \end{pmatrix}$

To get from B to C you need to go right by $\sqrt{3}$ and up by 1.

d $\overrightarrow{FC} = \begin{pmatrix} 0 \\ 4 \end{pmatrix}$

F is immediately below C. C is 1 above B, so F must be 1 below A, which means the distance from F to C is 4.

e $\overrightarrow{AE} = \begin{pmatrix} \sqrt{3} \\ 0 \end{pmatrix} \begin{pmatrix} 2\sqrt{3} \\ 0 \end{pmatrix}$

The vector $\overrightarrow{AE}$ is the same as the vector $\overrightarrow{BD}$. D is level with B, so you just need to look at how far D is to the right of B. To get from B to C you go right by $\sqrt{3}$, so you need to double this.

f $\begin{pmatrix} 0 \\ 2 \end{pmatrix} + \begin{pmatrix} \sqrt{3} \\ 1 \end{pmatrix} = \begin{pmatrix} \sqrt{3} \\ 3 \end{pmatrix}$

The vector $\overrightarrow{AC}$ is the same as $\overrightarrow{AB} + \overrightarrow{BC}$.

g $\overrightarrow{EF} = \begin{pmatrix} -\sqrt{3} \\ 1 \end{pmatrix}$

$\overrightarrow{EF}$ is the same vector as $\overrightarrow{CB}$. This is $\begin{pmatrix} -\sqrt{3} \\ 1 \end{pmatrix}$ (the same numbers as $\overrightarrow{BC}$ but with the opposite sign).

A well-labelled diagram might help.

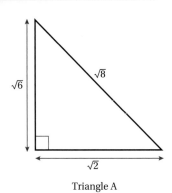

Triangle A

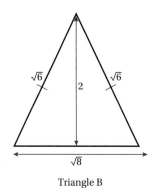

Triangle B

The base of the first triangle is 4 to the power of $\frac{1}{4}$, which is the same as $(4^{\frac{1}{2}})^{\frac{1}{2}}$

This is equal to $\sqrt{2}$.

In Triangle A the hypotenuse is $\sqrt{(6)^2 + (2)^2} = \sqrt{8}$

In Triangle B the two equal sides are $\sqrt{(8)^2 + (2)^2} = \sqrt{6}$

a The perimeter of Triangle A is $\sqrt{8} + \sqrt{6} + \sqrt{2}$

The perimeter of Triangle B is $\sqrt{8} + \sqrt{6} + \sqrt{6}$

Because $\sqrt{6} > \sqrt{2}$ Triangle B has the bigger perimeter.

b The area of Triangle A is

$\frac{1}{2}\sqrt{2}\sqrt{6} = \frac{1}{2}\sqrt{12} = \frac{1}{2}\sqrt{4}\sqrt{3} = \sqrt{3}$

The area of Triangle B is $\frac{1}{2}\sqrt{8} \times 2 = \sqrt{8}$

Because $\sqrt{8} > \sqrt{3}$ Triangle B has the bigger area.

21 **a**

You are told the radius of the circle is 1 cm, so it makes sense to draw in the radius.

The radius is 1 cm so the area of the shaded triangle is $\frac{1}{2}$ and the area of the whole square is 2 cm².

b

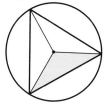

The shaded triangle has two sides of 1 cm and the angle between them is 120°.

The area, using $\frac{1}{2}ab\sin C$, is $\frac{1}{2}\sin 120°$.

The total area of the equilateral triangle is therefore 1.299.

The ratio of the area of the triangle to the area of the square is 1.299 : 2, which is approximately 2 : 3.

Divide the ratio by 2 then multiply by 3 to get 1.9485 : 3, which is approximately 2 : 3.

c

The shaded triangle has two sides of 1 cm with an angle between them of 72°.

The area, using $\frac{1}{2}ab\sin C$, is $\frac{1}{2}\sin 72°$.

The total area of the regular pentagon is therefore 2.3776...

The area of the pentagon divided by the area of the square is 1.18882... This is approximately 1.2, which is $\frac{6}{5}$.

22 **a**

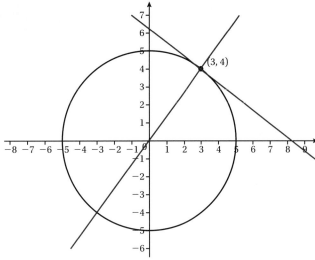

Add some extra lines to help.

The equation of the line passing through (0,0) is
$y = \frac{4}{3}x$

This means the equation of the tangent has a gradient of $-\frac{3}{4}$.

$y = -\frac{3}{4}x$ passes through the point (3, 4). Substituting in those values gives: $4 = -\frac{3}{4} \times 3 + c$, which rearranges to give $c = \frac{25}{4}$, which is $6\frac{1}{4}$.

This means the equation of the tangent is
$y = -\frac{3}{4}x + 6\frac{1}{4}$

The extra lines show that you have the radius and the tangent, so the gradients of the two lines multiply together to give −1.

b

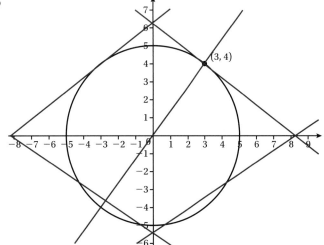

There are lots of other tangent lines that you could write down equations for, but the three shown here seem to be the easiest.

The original tangent was $y = -\frac{3}{4}x + 6\frac{1}{4}$

The tangent parallel to it is $y = -\frac{3}{4}x - 6\frac{1}{4}$

The reflection of the original tangent in the y-axis gives $y = -\frac{3}{4}x + 6\frac{1}{4}$

The reflection of the original tangent in the x-axis gives $y = -\frac{3}{4}x - 6\frac{1}{4}$

1

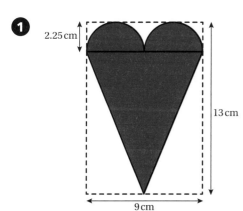

The two semicircles make a circle, radius 2.25 cm.

Area of circle = $\pi \times 2.25^2 = 15.9043 \ldots$ cm^2

Height of triangle = $13 - 2.25 = 10.75$ cm

Area of triangle = $\frac{1}{2} \times 9 \times 10.75 = 48.375$ cm^2

Area of red cardboard removed =
$9 \times 13 - 15.9043 - 48.375 = 52.7$ cm^2 (3 sf)

> Start by working out the areas of the different parts of the heart.

> Subtract the areas of the circle and triangle from the area of the rectangle.

2 **a** The diagram shows two wedges. That means 20 of these pairs total a metre long, so x must be 5 cm.

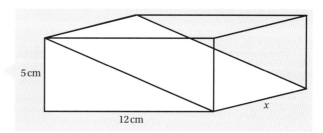

b To make £30 profit and cover the cost of the wood (£5), Ahmed has to take at least £35.

As he has 40 wedges to sell, the minimum cost to achieve this is £35 ÷ 40 = 87.5p for each wedge.

This rounds to 88p.

3 **a** Principal $\times 1.08^{10}$

$= \text{Principal} \times 2.16$

Yes, he would more than double his money.

8% interest is an increase with a multiplier of 1.08.

b $£40\,000 \times 1.08^{10}$

$= £86\,357$ to the nearest pound

c $£30\,000 \times 1.08^{10} = £64\,767.75$

$£31\,000 \times 1.08^{10} = £66\,926.67$

$£32\,000 \times 1.08^{10} = £69\,085.60$

$£33\,000 \times 1.08^{10} = £71\,244.52$

$£34\,000 \times 1.08^{10} = £73\,403.45$

$£35\,000 \times 1.08^{10} = £75\,572.37$

She must invest at least £35 000.

The bank only gives this deal for £30 000 or more so start checking from this amount.

Another way to work it out is to divide £75 000 by 1.08^{10}.

4 $1\,000\,000 \div 60 = 16\,666.666...$ minutes

$16\,666.666... \div 60 = 277.777...$ hours

$277.777... \div 24 = 11.574$ days

If each number takes about 2 seconds to say (on average), then it will take about 23 days to say them all.

This method starts by working out how many days are the same as 1 million seconds.

Some numbers will be much quicker to say than others. It is quick to say 'three' or 'five hundred thousand', but it takes much longer to say 'two hundred and twenty-two thousand, two hundred and twenty-two'.

Even allowing for breaks for sleeping and eating, you could easily do this in your lifetime.

5

a $4^3 = 4 \times 4 \times 4$, which is even

$3^4 = 3 \times 3 \times 3 \times 3$, which is odd

Even + odd = odd

So $4^3 + 3^4$ is odd

b 6^7 is even

3^7 is odd

So $6^7 + 3^7$ is odd

6 Probability Helen does not have to stop

$= \dfrac{9}{10} \times \dfrac{3}{4} \times \dfrac{1}{3} = \dfrac{27}{120}$

Probability Helen has to stop at least once

$= 1 - \dfrac{27}{120} = \dfrac{93}{120}$

This simplifies to $\dfrac{31}{40}$.

7 **a**

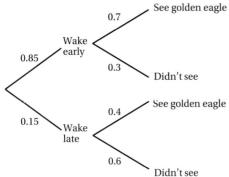

Probability he sees a golden eagle
$$= (0.85 \times 0.7) + (0.15 \times 0.4)$$

$$= 0.655$$

Hamish could see a golden eagle if he wakes up early, but he could also see one if he wakes up later.

b

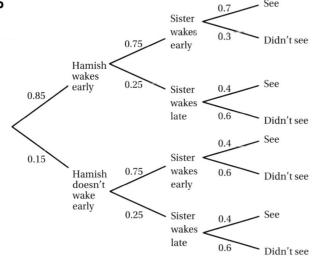

If Hamish wakes up late then it doesn't matter whether his sister wakes up early or late.

Probability they see golden eagle =
$(0.85 \times 0.75 \times 0.7) + (0.85 \times 0.25 \times 0.4) + (0.15 \times 0.4)$

$$= 0.59125$$

Probability they don't see golden eagle =
$1 - 0.59125 = 0.40875$

c The probability goes down from about 0.7 to about 0.6.

Without Elspeth, Hamish's chance of seeing a golden eagle is 0.655. With Elspeth, his chance of seeing a golden eagle is 0.591.

8　**a**　$140 \div 1\frac{1}{2} = 93.3\,\text{km/h}$

　　b　$140 \div 1\frac{1}{3} = 105\,\text{km/h}$

　　c　10 minutes

　　d　12 noon

　　e　4.5 hours, $\dfrac{280}{4.5} = 62.2\,\text{km/h}$

　　f　3 hours 40 minutes, $\dfrac{280}{3.666} = 76.36\,\text{km/h}$

　　g　105 km/h is only 65.6 miles per hour, so neither of them was speeding at any point as long as the speed limit was 70 mph.

> Read the graph to work out the times: Dave takes 1h 30 min to travel 140 km. Hazel takes 1 h 20 min to travel the first stage of her journey.

9　**a**

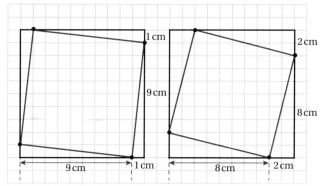

> Label what you know. Use the squares on the paper to find the lengths.

$a^2 + b^2 = c^2$　　　　　$a^2 + b^2 = c^2$

$9^2 + 1^2 = c^2$　　　　　$8^2 + 2^2 = c^2$

$82 = c^2$　　　　　　　$68 = c^2$

$c = \sqrt{82} = 9.055...\text{cm}$　　$c = \sqrt{68} = 8.246...\text{cm}$

> Now focus on a triangle. Use Pythagoras' theorem to calculate the side length of the titled square.

A line in the first logo is longer than a line in the second logo, so the total line length of the first logo is greater.

b

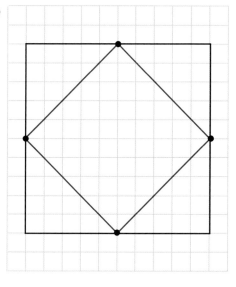

> To make a square of half the area you can rotate it.

> You know this has half the area because the diagonals of the inner square create 8 identical triangles, 4 of which make up the inner square.

10

It is sensible to multiply the number of years by 365 to work out the approximate number of days and to round off the final answer for the following reasons:

- You don't know what time of day they were born, so you don't know whether it is an exact number of days.

- You don't know which month to start with for Ross (some are longer than others).

- You also don't know how many leap years have been involved.

	Shirley	**Ross**	**Jemima**
		169 months	
Number of months / 12 = number of years	14.2 years	14.0833333 years	
Number of years × 365 = number of days	5183 days	5140.41667 days	5293 days
Number of days × 24 = number of hours	124392 hours	123370 hours	127032 hours
Number of hours × 60 = number of minutes	7463520 minutes	7402200 minutes	7621920 minutes
Number of minutes × 60 = number of seconds	447811200 seconds	444132000 seconds	457315200 seconds
Rounded to 3 sf	448000000 seconds	444000000 seconds	457000000 seconds

Jemima is the eldest (and is older than Dave by about 7 million seconds.

a

Work through the information step by step. A table might help.

	12 Muffins	30 Muffins	Fat	Sugar
Eggs	2	5	$4.6 \times 5 = 23$	trace 0
Caster sugar	200 g	500 g		500
Milk	250 ml	625 ml	$0.02 \times 625 = 12.5$	$0.05 \times 625 = 31.25$
Vegetable oil	125 ml	312.5 ml	312.5	0
Flour	400 g	1000 g	$7/500 \times 1000 = 14$	$1/1000 \times 1000 = 1$
Salt	1 tsp	2.5 tsp		
			total fat for 30 muffins: 362 g	total sugar for 30 muffins: 532.25 g
			fat for 1 muffin: 12.1 g	sugar for 1 muffin: 17.7 g

b Fat: $\dfrac{12.1}{70} = 17\%$ of the GDA

Sugar: $\dfrac{17.7}{90} = 20\%$ of the GDA

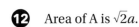

12 Area of A is $\sqrt{2}a$.

Start by working out what you know.

Area of B is $2\sqrt{5} \times \sqrt{10} = 2\sqrt{5} \times \sqrt{5} \times \sqrt{2} = 2 \times 5 \times \sqrt{2} = 10\sqrt{2}$

The areas of the two rectangles are the same, so $\sqrt{2}a = 10\sqrt{2}$ and it is clear that $a = 10$.

13 $\sqrt{28} \times \sqrt{12}$

Start by working out the area of the bigger rectangle.

$\sqrt{28} \times \sqrt{12} = \sqrt{4} \times \sqrt{7} \times \sqrt{4} \times \sqrt{3} = 4\sqrt{21}$

It's easier to simplify this now.

$\sqrt{7} \times \sqrt{3} = \sqrt{21}$

Now work out the area of the smaller rectangle.

The fraction that is left is $\dfrac{3\sqrt{21}}{4\sqrt{21}} = \dfrac{3}{4}$

14 $V_{\text{hemisphere}}$ is $\dfrac{4}{3}\pi r^3 \div 2 = \dfrac{2}{3}\pi r^3$

Work out the volume of a hemisphere.

V_{cone} is $\dfrac{1}{3}\pi r^2 h$

Work out the volume of a cone.

$V_{\text{cone}} = \dfrac{1}{3}\pi r^2 \times 2r = \dfrac{2}{3}\pi r^3$

The conical hanging basket has a height equal to its diameter. Therefore $h = 2r$.

The volumes are identical.

15

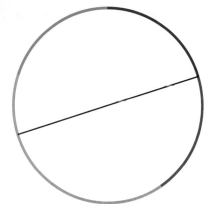

Five of the squares fit around the mug, so start with five lines of the same length around the mug.

The grey line is the diameter (9 cm).

The circumference of the circle is 9π, so each coloured line (representing a diagonal of one of the squares) is $\frac{9\pi}{5} = 5.654867\ldots$

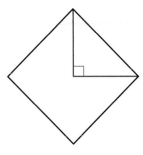

The red lines on this square are half of 5.654867....

$\sqrt{2.8\ldots^2 + 2.8\ldots^2} = 3.998\ldots$

The perimeter is four times this:
$15.99438 = 16.0 \text{ cm} (3 \text{ sf}).$

You can use Pythagoras' theorem to work out the hypotenuse of the right-angled triangle.

1 **a** For example, $\frac{1}{50}, \frac{1}{49}, \frac{1}{48}, \frac{1}{47}, \frac{1}{46}$

This question can be answered in many ways, but making changes to the original set of fractions works well.

b For example, $\frac{2}{6}, \frac{2}{5}, \frac{2}{4}, \frac{2}{3}, \frac{2}{2}$

c For example, $\frac{6}{2}, \frac{7}{2}, \frac{8}{2}, \frac{9}{2}, \frac{10}{2}$

or

$\frac{5}{3}, \frac{7}{4}, \frac{9}{5}, \frac{8}{3}, \frac{9}{2}$

You could sprinkle some unrelated fractions around (such as in this final example), but it is easier to work in a systematic way and to make use of some patterns.

---

2 **a** A diagram shows that 12 chairs are required.

Every table has 2 chairs, and there are 2 extra on the ends (shown in red).

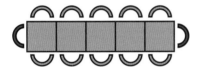

b The rule is $2n + 2$ chairs, where n is the number of tables.

A diagram shows the structure of the algebra.

c If it is possible to use 55 chairs then $2n + 2 = 55$, so $n = 26.5$. You can't have half a table, so this is not possible.

d Each big group will contain 16 people. $2n + 2$ means that $n = 7$, so each group will require 7 tables.

This just involves a tweak to the rule from part **b**.

---

3 **a** $\frac{3}{7}$ of the balls in the bag = 9 balls

Consider what you know.

$\frac{1}{7}$ of the balls in the bag = 3 balls

The number of balls in the bag = $3 \times 7 = 21$ balls

b $\frac{3}{25}$ of the balls in the bag = 6 green balls

$\frac{1}{25}$ of the balls in the bag = 2 balls

The number of balls in the bag = $2 \times 25 = 50$ balls

a $400\,\text{mg} = 0.4\,\text{g}$
$400\,\text{g} + 400\,\text{mg} = 400.4\,\text{g}$

$0.5\,\text{kg} = 500\,\text{g}$
$0.5\,\text{kg} - 90\,\text{g} = 410\,\text{g}$

$400\,\text{g} + 400\,\text{mg} < 0.5\,\text{kg} - 90\,\text{g}$

> Change the quantities so they use the same units. It makes sense to work in grams. $1\,\text{kg} = 1000\,\text{g}$ and $1\,\text{g} = 1000\,\text{mg}$.

b $0.1\,\text{km} = 100\,\text{m}$
$150\,\text{cm} = 1.5\,\text{m}$
$0.1\,\text{km} + 150\,\text{cm} = 101.5\,\text{m}$

$900\,\text{cm} = 9\,\text{m}$
$110\,\text{m} - 900\,\text{cm} = 101\,\text{m}$

$0.1\,\text{km} + 150\,\text{cm} > 110\,\text{m} - 900\,\text{cm}$

> Convert everything to metres. $1\,\text{km} = 1000\,\text{m}$ and $1\,\text{m} = 100\,\text{cm}$.

c $0.75\,\text{hours} = 45\,\text{minutes}$
$600\,\text{seconds} = 10\,\text{minutes}$
$0.75\,\text{hours} + 600\,\text{seconds} = 55\,\text{minutes}$

$0.1\,\text{hours} = 6\,\text{minutes}$
$50\,\text{minutes} + 0.1\,\text{hours} = 56\,\text{minutes}$

$0.75\,\text{hours} + 600\,\text{seconds} < 50\,\text{minutes} + 0.1\,\text{hours}$

> Convert everything to minutes.

> Consider the numbers in the question. What happens if you change any of them? The rate of growth of the hair doesn't make a difference to the number of hairs that are lost. The number of strands of hair that people start with doesn't affect the proportion that they lose. So you just need to look at the proportion that is lost each year.

a Raise 0.9975 to the power of 82 to find out the proportion that will be left after 82 years.

> If you lose 10% of something you have 90% left. In this case the average person loses 0.25% of their hair per year so they keep 99.75% per year.

$0.9975^{82} = 0.814438$ so about 81% remains, which means about 19% has been lost.

b $0.9975^{50} = 0.882359$

$0.9975^{40} = 0.904724$

$0.9975^{42} = 0.900206$

$0.9975^{43} = 0.8979556$

It would take 43 years for at least 10% of the hair to be lost.

> If you lose 10% then you have 90% left, which is 0.9. You need to know what n is when $0.9975^{n} \leq 0.9$.
>
> You can use trial and improvement.

6

$24 = 4 \times 6$, so $\sqrt{24} = \sqrt{4} \times \sqrt{6} = 2\sqrt{6}$

$54 = 9 \times 6$, so $\sqrt{54} = \sqrt{9} \times \sqrt{6} = 3\sqrt{6}$

Both of these involve 6, so check whether 96 is divisible by 6.

$96 \div 6 = 16$

$96 = 16 \times 6$, so $\sqrt{96} = \sqrt{16} \times \sqrt{6} = 4\sqrt{6}$

This means $\dfrac{\sqrt{24} + \sqrt{54} + \sqrt{96}}{3} = \dfrac{2\sqrt{6} + 3\sqrt{6} + 4\sqrt{6}}{3} = \dfrac{9\sqrt{6}}{3}$

$$= 3\sqrt{6}$$

To work out the mean of three numbers, add them together and divide by three. As this is a non-calculator question you will need to combine the surds by hand.

To simplify surds, look for square numbers.

7 **a** $1 + 2 + 3 = 6$, which is 2×3

$2 + 3 + 4 = 9$, which is 3×3

$3 + 4 + 5 = 12$, which is 4×3

First, try it out with some more sets of three consecutive integers. You could work systematically.

Once you are convinced that it works, try to think about why.

$(n - 1) + (n) + (n + 1)$

$n - 1 + n + n + 1 = 3n$

Yes, this is true for all integers.

There are several different ways of saying how this works. To start with, think about a general set of three consecutive numbers.

Using algebra you can simplify this expression.

The expression simplifies to $3n$ ($3 \times$ the middle number). So it must be true for all integers.

b $n + (n + 1) + (n + 2) + (n + 3) = 4n + 6$

You could do this algebraically, where n is the first number.

c

Number of numbers	Rule based on the first number being n
2	$2n + 1$
3	$3n + 3$
4	$4n + 6$
5	$5n + 10$
6	$6n + 15$
7	$7n + 21$
8	$8n + 28$

The method for part **b** can be extended for any number of numbers.

8 **a** Circular pens: $C = \pi d$

Fences with 2 m gap between:

Inner fence, diameter = 20 m, $C = 20\pi$

Outer fence, diameter = 24 m, $C = 24\pi$

Difference = 4π

> The formula for the circumference C of a circle with diameter d is $C = \pi d$.

b Fences with 3 m gap between:

Outer fence, diameter = 26 m, $C = 26\pi$

Difference = 6π

> Now work out the difference for a 3 m gap between the fences.

c

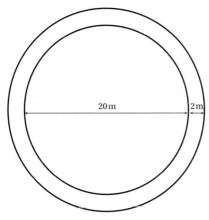

The outer fence, diameter = $20 + 2x$, $C = (20 + 2x)\pi = 20\pi + 2\pi x$

The inner fence circumference is 20π, so for a gap of x the difference in the length of the fences is $2\pi x$.

d

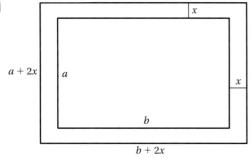

> This diagram shows what happens for any rectangles, where the inner one is a by b and there is a gap of x all the way around.

The perimeter of the inner rectangle is $a + b + a + b = 2a + 2b$

The perimeter of the outer rectangle is $(a + 2x) + (b + 2x) + (a + 2x) + (b + 2x) = 2a + 2b + 8x$

The difference between these is $8x$.

9 To 1 significant figure this calculation is:

20 × 800 = 16 000, which is about 10 times the size of 1903.02. The answer is likely to be incorrect.

Start with 23.64 × 805 = 1903.02

If 23.64 × 805 = 1903.02 is true, then:

2.364 × 805 = 190.302

and 2.364 × 8.05 = 1.90302

The left-hand side must be bigger than 16, so the calculation can't be correct.

> Here is one convincing method that involves making changes to the problem.

> Here is a second convincing method that involves making different changes.

---

10 Snail 1: 36 000 mm/h (÷ 10)

3600 cm/h (÷ 60)

60 cm/min (÷ 60)

1 cm/s

Snail 2: 0.01 m/s (× 100)

1 cm/s

Snail 3: 5 km/day (× 1000)

5000 m/day (× 100)

500 000 cm/day (÷ 24)

20 833.333... cm/h (÷ 60)

347.222... cm/min (÷ 60)

5.787... cm/s (= 5.8 cm/s to 1 dp)

Snail 4: 700 cm/h (÷ 60)

11.666... cm/min (÷ 60)

0.19444... cm/s (= 0.2 cm/s to 1 dp)

Snail 3 finishes first (and is a fast snail!).

> To compare the snails' speeds they must be measured in the same units.
>
> In this question it might be sensible to convert all the speeds into cm per second (cm/s).

11 Volume of Choc Flakes packet
= 28 cm × 17 cm × 5 cm = 2380 cm³

Volume of BioWheat packet
= 21 cm × 15 cm × 9 cm = 2835 cm³

The volume of the BioWheat packet is greater than the volume of the Choc Flakes packet, therefore Mary is wrong.

Surface area of Choc Flakes packet
= 2 × (28 × 17) + 2 × (28 × 5) + 2 × (17 × 5) = 1402 cm²

Surface area of BioWheat packet
= 2 × (21 × 15) + 2 × (21 × 9) + 2 × (15 × 9) = 1278 cm²

The surface area of the Choc Flakes packet is greater than the surface area of the BioWheat packet, therefore Kelly is correct.

Raymon is wrong as the packet with the greater volume does not have the greater surface area.

> Work out the volume and surface area of each packet.

> Conclusion part 1.

> Conclusion part 2.

> Conclusion part 3.

---

 12 **a** Daisy can move her shape up and down.

A translation using any vector $\begin{pmatrix} 0 \\ a \end{pmatrix}$ will do this.

> This is a translation where the x-coordinate doesn't change but the y-coordinate can change to be anything.

b A translation using any vector $\begin{pmatrix} b \\ b \end{pmatrix}$ will do this. This can also be written as $b\begin{pmatrix} 1 \\ 1 \end{pmatrix}$.

> This time Daisy moves her shape diagonally. If she moves it right then she also has to move it up by the same amount.

c

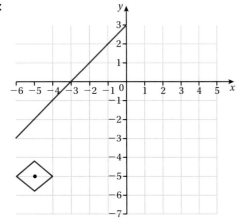

> A good diagram will help here.

To get the black dot onto this line you could use the vector $\begin{pmatrix} 0 \\ 3 \end{pmatrix}$ or $\begin{pmatrix} 1 \\ 4 \end{pmatrix}$ or $\begin{pmatrix} 2 \\ 5 \end{pmatrix}$.
You always need to go up three more than you go across, so the vector $\begin{pmatrix} c \\ c+3 \end{pmatrix}$ will cover all the possibilities (including those with decimals).

a 1 Try a different starting shape.

2 Try a different pair of perpendicular mirror lines.

3 Try a different starting position for the shape.

b

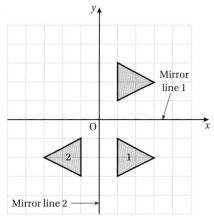

Mikey has looked at one example and drawn his conclusion based on this.

If you carry out the reflection in Mikey's chosen mirror lines ($y = 0$ and $x = 0$) with this new shape it becomes clear that something else is happening: the shape has been rotated 180° about the origin.

Because Mikey's original shape was a rectangle, it has rotational symmetry and he therefore did not notice that the rotation had happened.

His conclusion was not incorrect, but it was not the full answer. That depends on the nature of the starting shape.

 The perimeter of the large semicircle is

$$\pi \times \frac{6}{2} = 9.42\ldots \text{cm}$$

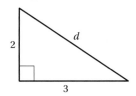

Treat the perimeter as a large semicircle and a smaller circle.

Now find the diameter of the smaller circle.

This is the hypotenuse of a right-angled triangle.

$a^2 = 2^2 + 3^2$

$d = \sqrt{13} = 3.60555\ldots \text{cm}$

Circumference $= \pi \times 3.60655\ldots \text{cm} = 11.327\ldots \text{cm}$

Total perimeter $= 9.42478\ldots + 11.327\ldots = 20.75 \text{cm} (2 \text{ dp})$

15 a

Fraction	Decimal
$\dfrac{1}{9}$	0.1111...
$\dfrac{1}{99}$	0.01010...
$\dfrac{1}{999}$	0.001001001...
$\dfrac{1}{9999}$	0.000100010001...
$\dfrac{1}{99999}$	0.00001000010000100001...
$\dfrac{1}{999999}$	0.000001000001...

It seems sensible to use a calculator and to work out the decimal equivalents.

All of the fractions have a denominator that is a multiple of 9.

All of the fractions have equivalent decimals that are eventually recurring.

The recurring part of each decimal includes only 1s and 0s.

The number of recurring digits (the period) is the same as the number of digits in the denominator.

Similarities

The denominator of each fraction is different. In fact the denominator is getting larger with each fraction.

The equivalent decimal of each fraction is smaller than the previous one.

The number of zeros in the recurring part of the fraction increases with the denominator.

Differences

b i $\dfrac{1}{9} = 0.1111...$

Therefore $0.7777... = \dfrac{7}{9}$

Look for similarities between the number and the fractions considered above.

This is similar to the recurring decimal for $\dfrac{1}{9}$. In fact it is seven times bigger.

ii $\dfrac{1}{9} = 0.1111...$

$0.333... = \dfrac{3}{9} = \dfrac{1}{3}$

Therefore $5.3333... = 5\dfrac{1}{3}$

This is also equal to $\dfrac{16}{3}$

iii $0.14444...$ is the same as $0.44444... - 0.3$

$0.44444... = \dfrac{4}{9}$ so you need to do $\dfrac{4}{9} - \dfrac{3}{10}$

This is $\dfrac{40}{90} - \dfrac{27}{90} = \dfrac{13}{90}$

iv $2.5000900090009...$ is

$2.5 + (0.000100010001...) \times 9 \div 10$

This is $2\dfrac{1}{2} + \dfrac{9}{99\,990} = 2\dfrac{1}{2} + \dfrac{1}{11\,110}$

$= 2\dfrac{5555}{11\,110} + \dfrac{1}{11\,110} = 2\dfrac{5556}{11\,110} = 2\dfrac{2778}{5555}$

16 **a**

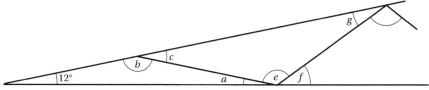

It is possible to work out every angle in the diagram. Start on the left hand side of the diagram.

Work out angle *a*.

Base angles in an isosceles triangle are equal.

Work out angle *b*.

Angles in a triangle add up to 180°.

Work out angle *c*.

Angles on a straight line add up to 180°.

Work out angle *d*.

Base angles in an isosceles triangle are equal etc

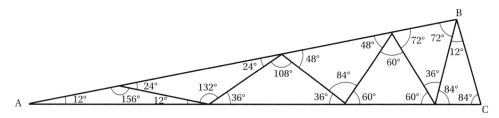

Check whether the angles at ABC and ACB are equal.

This means the outer triangle is isosceles.

So Helen is correct.

b Repeating with a starting angle of 10° gives the following triangle.

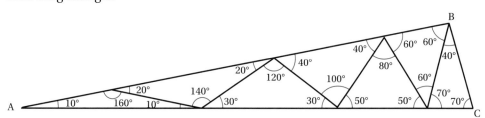

Pete's suggestion of starting with an angle of 10° does not result in an outer isosceles triangle since angle ABC ≠ angle ACB.

c

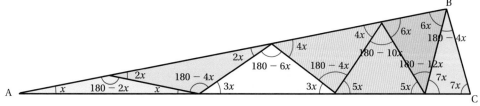

This is an algebraic solution, starting with angle *x*.

Angle ABC = angle ACB (because this is an isosceles triangle).

This means that $6x + y = 7x$, so $x = y$.

The three angles in triangle ABC add up to $15x$, so $15x = 180°$.

This means that the only starting value that works is 12°.

17

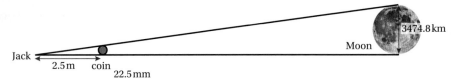

Draw a diagram to illustrate the situation. This is a problem about similar triangles (the triangle from Jack to the coin and the triangle from Jack to the moon).

The question says Jack is about 2.5 m away from the coin, so it seems sensible to round off the value for the diameter of the Moon to 3500 km.

3500 km = 3 500 000 m = 350 000 000 cm

22.5 mm = 2.25 cm

Calculate the scale factor of enlargement from the coin diameter to the Moon diameter.

Scale factor = 350 000 000 cm ÷ 2.25 = 155 555 555

2.5 m × 155 555 555 = 388 888 889 m

Use the scale factor to enlarge the distance.

Distance to the Moon = 390 000 km (rounded to 2 sf)

1

| | | $\frac{1}{4}$ of the bags | | $\frac{1}{4}$ of the bags | | $\frac{1}{4}$ of the bags | | $\frac{1}{4}$ of the bags | | | | Supermarket A |

| | $\frac{1}{4}$ of the bags | $\frac{1}{4}$ of the bags | | | $\frac{1}{4}$ of the bags | | | $\frac{1}{4}$ of the bags | | Supermarket B |

5   6   7   8   9   10   11   12   13   14

> You don't have enough information to draw a complete box plot, but you can draw the parts for which you do have information.

If there are more apples in the bags then, because all the bags weigh 1 kg, each apple must weigh less. The biggest apples will presumably be in the bags that have fewer apples in them.

For Supermarket B half the bags have 8 apples or fewer, whereas at Supermarket A half the bags have 9 apples or fewer. This suggests that the bigger apples are at Supermarket B.

But $\frac{1}{4}$ of the bags at Supermarket A have 6 apples or fewer, which is better than at Supermarket B where $\frac{1}{4}$ of the bags have 7 apples or fewer, so this suggests that the bigger apples are at Supermarket A.

2

Husband		Son
8	57	
9 8 5	58	1 7 8 9 9
5 5	59	1 3 3 4
6 1	60	2 4
3 1	61	
3	62	

Key: 57 | 4 = 57.4 g

> A stem-and-leaf diagram would be a good way to display the data.

This shows that there are more packets that are less than 60 g than are 60 g or more, so the farmer is wrong to label the packets 60 g.

The median is 59.5 g for her husband and 59.1 g for her son.

It might be better to label the packets as being 59 g instead.

3　**a**　$12\,m \times 8\,m = 96\,m^2$, so the cost is $£4 \times 96 = £384$.

> The first lawn has an area of $80\,m^2$ and this costs £320, so each square metre costs £4.
>
> The second lawn has an area of $144\,m^2$ and, as expected, the cost is $£4 \times 144$.

　　b　$560 \div 4 = 140$, so the area is $140\,m^2$.

　　　　Any pair of integers that multiply to give 140 will work: 1×140, 2×70, 4×35, 5×28, 7×20, 10×14.

4　It might be sensible to assume a working day of 8 hours (9 am – 5 pm).

> Clearly this is a joke, but assume it is accurate. You need to make some assumptions about Nick's 'normal' working hours.

　　a　A full working day would be 8×60 minutes = 480 minutes. 5% of this is 24 minutes, so Nick would work 24 minutes on a Friday.

> Calculate the amount of time spent working on Friday. 5% of 8 hours is 0.4 of an hour. This doesn't mean very much, so change the 8 hours into minutes.

　　b　On Wednesday Nick would work 40% of 480 mins (192 mins). On Monday he would work 12% of 480 mins (57 mins and 36 seconds). Therefore, subtracting, Nick would work 134 mins and 24 seconds (2 hours 14 mins 24 seconds) more on a Wednesday.

　　c　It might look as if it must be Nick because he works for 23% compared to Bernard's 18%, but if Bernard's working hours are longer then 18% of a larger amount could be bigger than 23% of something smaller.

5　**a**　225 minutes = 3 hours 45 minutes

> A lot of information in the question is irrelevant. The carving time is the same for both birds. The turkey weighs 4.5 kg more, so it needs to be in the oven for 50 minutes × 4.5 longer.

　　b　$t = 50m + 30$

> Convert the time to minutes then substitute in the equation from part **b**.

　　c　$5 \times 60 + 5 = 305$ minutes

　　　　$(305 - 30) \div 50 = 5.5\,kg$

　　d　9.25 am if no carving time is required.

　　　　9.05 am if 20 minutes of carving time is needed.

6

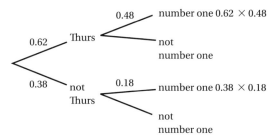

This type of question can be solved using a tree diagram.

The probability of a number one hit is 0.366.

7 **a** +1, –1

Do you recognise the equations of perpendicular graphs?

b $y = -\frac{1}{2}x - 3$

c The product of the gradients of two perpendicular lines is –1.

8

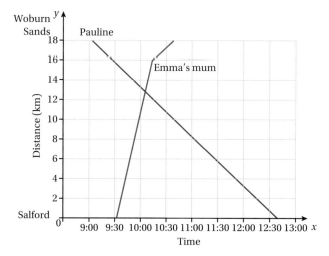

It is useful to recognise that a travel graph is likely to help here.

a 10:10

b 6 km

c 12 km

d 12:10

e 6 km/h

f 5 km

173

9　**a**　10 km

　　b　20 minutes

　　c　2 stops, 20 minutes

　　d　First stage of the return journey.

　　e　30 km in 10 minutes, means 180 km/h

　　f　20:55

　　g　about 37 km

　　h　90 km ÷ 3 h = 30 km/h

　　i　90 km/h

Use the travel graph to help you answer the questions.

The graph is steepest at this point.

10　$f(x) + 7 = 2x + 4 + 7 = 2x + 11$, which has y-intercept of 11.

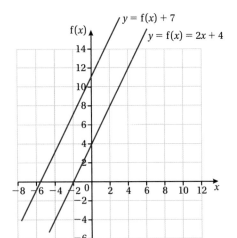

You could give reasons for this using algebra.

Alternatively, you could draw a graph.

11 **a** You would expect a weak negative correlation.

> It is likely that Year 7 students go to bed earlier and therefore have more sleep than Year 11 students.

b Here is an example:

"Year 11 student have less sleep than Year 7 students."

c An example:

Question 1: "What year group are you in?"

Tick one box:

Yr 7 ☐ Yr 8 ☐ Yr 9 ☐ Yr 10 ☐ Yr 11 ☐

Question 2: "On average, how many hours of sleep do you have per night?"

Tick one box.

$4h < s \leqslant 5h$ ☐

$5h < s \leqslant 6h$ ☐

$6h < s \leqslant 7h$ ☐

$7h < s \leqslant 8h$ ☐

$8+h$ ☐

d A scatter diagram will show immediately if there is any correlation and possible outliers.

---

a All the quickest journeys involve one of the edges of each length: $4.5\,m + 3\,m + 2\,m = 9.5\,m$.

b $GD = \sqrt{4.5^2 + 2^2} = 4.92\,m$

> Find GD using Pythagoras' theorem.

From D to A = 3 m

Total distance walked = 4.92 m + 3 m = 7.92 m

This is 9.5 – 7.92 = 1.58 m shorter.

> Don't forget to work out the difference between the two path lengths as this is what was asked for in the question.

13

The biggest difference between the areas must happen with the upper bound.

The upper bound of 7.0 is 7.05 cm.

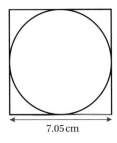

Think about the area between the square and the circle. As the square and circle both get bigger, the area will get bigger (if this is not clear you could imagine a very big square/circle and a very small square/circle).

7.05 cm

$7.05 \times 7.05 = 49.7025 \text{ cm}^2$

Calculate the upper bound for the area of the square.

$$\pi \times r^2 = \pi \times \left(\frac{7.05}{2}\right)^2 = 39.036... \text{ cm}^2$$

Calculate the area of the circle for the upper bound of the square side length.

Area of square – area of circle

Calculate the difference between the two areas.

$= 49.7025 \text{ cm}^2 - 39.036... \text{ cm}^2 = 10.666... \text{ cm}^2$

The maximum difference between the area of the square and the area of the circle is 10.7 cm^2 (to 3 sf).

---

14 **a** 42, 56, 72

The sequence numbers have factors. For example:

0	2	6	12	20	30
0×1	1×2	2×3	3×4	4×5	5×6

b $(n-1) \times n$

This can be written as $n(n-1)$, or as $n^2 - n$.

Each term is a product of two adjacent numbers. The next three numbers will be 6×7, 7×8, 8×9.

c $n(n-1)$ is always a number multiplied by the previous number. One of those two numbers must be even, so the answer must be even too.

15 Radius of small bottle = 8.5 ÷ 2 = 4.25 cm

Volume of small bottle (cylinder) = area of circular base × height = $\pi \times 4.25^2 \times 23 = 1305.14$ cm^3 (2 dp)

Volume of large bottle = $\pi \times 5.5^2 \times 25 = 2375.83$ cm^3 (2 dp)

Difference between the volumes of the large and small bottles = 2375.83 – 1305.14 = 1070.69 cm^3 (2 dp)

To the nearest ml, the large bottle holds 1071 ml or 1.071 litres more than the smaller bottle.

> Work out the volumes of the cylinders.

> Conclusion.

16 **a** For the red bricks: $2t + 2c \leq 8$

b For the yellow bricks: $2t + c \leq 6$

c

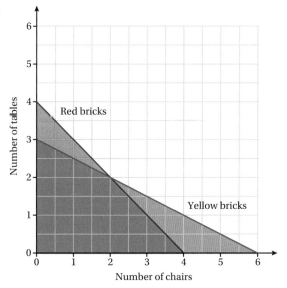

> Let t be the number of tables and c be the number of chairs.

17 $10\sqrt{15} = \sqrt{10^2}\sqrt{15} = \sqrt{100 \times 15} = \sqrt{1500}$

B is correct.

> $\sqrt{15}$ is a surd that cannot be simplified, so $10\sqrt{15}$ cannot possibly be an integer. This means C and D must be wrong.

18 a $\overrightarrow{AB} + \overrightarrow{BC} = \overrightarrow{AC}$

$\overrightarrow{AC} = 2\mathbf{a} - \mathbf{b} - (\mathbf{a} + 2\mathbf{b})$

$= \mathbf{a} - 3\mathbf{b}$

b $\overrightarrow{A_1B_1} = 3(2\mathbf{a} - \mathbf{b}) = 6\mathbf{a} - 3\mathbf{b}$

$\overrightarrow{CB} = \mathbf{a} + 2\mathbf{b}$ and $C_1B_1 = 3\mathbf{a} + 6\mathbf{b}$, so the lengths in triangle $A_1B_1C_1$ are 3 times bigger than the lengths in triangle ABC.

Hence A_1B_1 must be 3 times bigger than $\overrightarrow{AB}$.

19 a

In order to calculate the aspect ratio you need to know the height of Carlos's TV.

You know the diagonal length and the width so you need to apply Pythagoras' theorem.

Height = 18 inches

Aspect ratio = 32 : 18

This simplifies to 16 : 9.

b

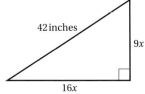

The aspect ratio of the TV is 16 : 9 (width : height).

You know the diagonal length so apply Pythagoras' theorem.

$(9x)^2 + (16x)^2 = 1764$

$337x^2 = 1764$

$x^2 = 1764 \div 337$

$x = \sqrt{1764 \div 337} = 2.287\ldots$ inches

Width = $16x = 16 \times 2.287\ldots = 36.6$ inches (to 1 dp)

Height = $9x = 9 \times 2.287\ldots = 20.6$ inches (to 1 dp)

20 **a**

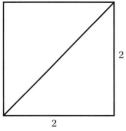

Diagonal $= \sqrt{2^2 + 2^2} = \sqrt{8}$

Use Pythagoras' theorem.

$\sqrt{8} = 2\sqrt{2}$

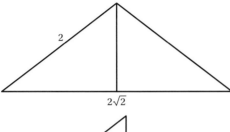

Now use that as the base of an isosceles triangle.

There are lots of ways to do this. One is to work out the length of the diagonal of the square base and then to use this with the slanted height.

Height $=$

$\sqrt{2^2 + (\sqrt{2})^2} = \sqrt{4 - 2} = \sqrt{2}$

The height of this triangle is the height of the pyramid. You need to halve the diagonal to get the base of the triangle.

b Height $= \dfrac{3}{2}\sqrt{2}$

You could repeat all the calculations from part **a**, but an alternative is to note that all of the lengths will change in the same way. To turn the 2-pyramid into a 3-pyramid you need to multiply by $\dfrac{3}{2}$.

c Height $= \dfrac{a}{2}\sqrt{2}$

To turn the 2-pyramid into an a-pyramid you need to multiply by $\dfrac{a}{2}$.

21

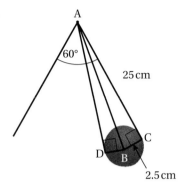

A

60°

25 cm

C

D B

2.5 cm

Angle ACB is a right angle (because the tongs form a tangent to the circle).

$\tan \text{BAC} = \dfrac{2.5}{25}$

Angle BAC = 5.710593°

The angle made by the tongs is double that: 11.421186°

The angle has been reduced by 60° – 11.421186° = 49° (to the nearest integer).

> Draw in some lines and see whether that helps. Then it is clear that there is a right-angled triangle and you can work out some angles.

---

22 $C = \pi \times \text{diameter}$

Radius = 3.2 cm = 32 mm

$C = \pi \times (32 \times 2) = 64\pi \,\text{mm}$

$121 \div 64\pi = 0.60 \text{ (to 2 dp)}$

The dial has been turned through 0.6 of a turn, which is equivalent to $\dfrac{3}{5}$ of a turn.

This means it ends up pointing at the Wool setting.

> Circumference (C) of dial is πd.

> Write the arc length as a fraction of the circumference.

---

23 **a** Upper bound = 1.5 × 2.5 × 3.5 = 13.125 m³

Lower bound = 0.5 × 1.5 × 2.5 = 1.875 m³

Difference is 11.25 m³

b Upper bound = 1.005 × 2.005 × 3.005 = 6.05515 m³

Lower bound = 0.995 × 1.995 × 2.995 = 5.97204 m³

Difference is 0.083 m³

c Upper bound = 1.0005 × 2.0005 × 3.0005
= 6.0055015 m³

Lower bound = 0.9995 × 1.9995 × 2.9995
= 5.9945015 m³

Difference is 0.011 m³

> The biggest volume is calculated using the three upper bounds and the smallest is calculated using the lower bounds.

---

24 a Shop A: Final percentage price $\frac{209}{310} \times 100 = 67.4\%$ of original price

There are lots of ways to work this out, including by calculating the original price.

Shop A offers a 32.6% discount (3 sf)

Shop B: Final percentage price $= \frac{259}{310} \times 100 = 83.5\%$ of original price

Shop B offers a 16.5% discount (3 sf)

b Shop A: Original price of trainers $= \frac{62}{67.4} \times 100 = £91.99$

Shop B: Final price $= 91.99 \times 0.835 = £76.81$

The trainers will cost £76.81 in Shop B.

---

 25 a $\frac{3}{\sqrt{5}} = \frac{6}{2\sqrt{5}}$

$\frac{6}{2\sqrt{5}} > \frac{5}{2\sqrt{5}}$

so $\frac{3}{\sqrt{5}}$ is larger.

In order to compare fractions they need to have the same denominator.

b $\sqrt{8} = \sqrt{(4 \times 2)} = \sqrt{4} \times \sqrt{2} = 2\sqrt{2}$

$\sqrt{50} = \sqrt{(25 \times 2)} = \sqrt{25} \times \sqrt{2} = 5\sqrt{2}$

Again you need to have the same denominator. It isn't easy to see how to do this for this question. First you could see if you can simplify the surds.

$\frac{4}{\sqrt{8}} = \frac{4}{2\sqrt{2}} = \frac{2}{\sqrt{2}}$

$\frac{25}{\sqrt{50}} = \frac{25}{5\sqrt{2}} = \frac{5}{\sqrt{2}}$

$\frac{2}{\sqrt{2}} < \frac{5}{\sqrt{2}}$

so $\frac{25}{\sqrt{50}}$ is larger.

Now you can see that to find a common denominator you can multiply the top and bottom of the first fraction by 5 and the top and bottom of the second fraction by 2.

c $\frac{10}{\sqrt{7} + 3}$ is a positive number (because the numerator and the denominator are both positive), but $\frac{6}{\sqrt{7} - 3}$ is a negative number (because the numerator is positive but the denominator is negative), so $\frac{10}{\sqrt{7} + 3}$ is larger.

This could be worked out by finding a common denominator for the two fractions like in parts **a** and **b**. An alternative way is shown.

1 **a** Coach Cooksey will have the more accurate mean height as he is using the actual height of each of his players. Coach McKay is finding the estimated mean using the midpoints of each class interval. By using the midpoints he is giving each player in each class interval the same height.

Team McKay			
Height (cm)	**Midpoint**	**Frequency**	**Midpoint × frequency**
$145 \leqslant h < 155$	150	1	150
$155 \leqslant h < 165$	160	2	320
$165 \leqslant h < 175$	170	2	340
$175 \leqslant h < 185$	180	8	1440
$185 \leqslant h < 195$	190	3	570
$195 \leqslant h < 225$	210	4	840
Totals:		20	3660

b $3660 \div 20 = 183$ cm

c The mean height is the same for both teams at 183 cm.

But Team Cooksey has both the shortest player (height 142 cm) and the tallest player (height 226 cm), so the team has a larger range of heights.

2 **a** $1500 \text{ m} \div 200 \text{ m} = 7.5$

Simon will need to run 7.5 laps of the indoor running track and 3.75 laps of the outdoor running track to cover 1500 m.

Given that the other track is twice as long, he will need half as many laps.

b $5 \text{ km} = 5000 \text{ m}$

$5000 \text{ m} \div 200 \text{ m} = 25$

Denise will have to run 25 laps of the indoor track and 12.5 laps of the outdoor track to cover 5 km.

The 10 km is double the 5 km race, so Denise will have to run 50 laps of the indoor track and 25 laps of the outdoor track to cover 10 km.

3 **a** $t = 40k + 25$, where t is the number of minutes and k is the number of kilograms.

b No, he will need to put it in the oven at 9:25 am.

$40 \times 3.75 + 25$ plus an extra 10 minutes of time is 185 minutes. This is the same as 3 hours and 5 minutes.

4 **a**

		Day 1		
		Swim	**Jog**	**Cycle**
Day 2	**Swim**	SS	JS	CS
	Jog	SJ	JJ	CJ
	Cycle	SC	JC	CC

This is one way to list all the possibilities in a systematic way.

b $\frac{1}{9}$

c $\frac{6}{9}$

$= \frac{2}{3}$

All of the combinations except SS, JJ and CC.

This is sensible because, whichever one is chosen for Day 1, there are 2 out of the 3 that can be chosen on Day 2.

---

5 **a** In the original recipe there is 150 g + 120 g = 270 g of dark chocolate.

$270\,\text{g} \times 2.5 = 675\,\text{g}$

You need to convert a recipe for 8 people to a recipe that serves 20 people.

You could divide by 8 to find the ingredients for one serving and then multiply by 20.

Alternatively, you could multiply by $\frac{20}{8}$, which is the same as multiplying by $2\frac{1}{2}$.

b 4 bars of 200 g each (because 3 bars is only 600 g)

c $800\,\text{g} - 675\,\text{g} = 125\,\text{g}$

675 : 125

27 : 5

d $\frac{3}{4}$ of 400 g is 300 g.

Fiona needs 300 g of chocolate spread.

To serve 6 people you can divide by 8 and multiply by 6, or you can multiply by $\frac{6}{8}$, which is the same as $\frac{3}{4}$.

---

6 **a** Each cleaner works for 2 hours 45 mins, which is 165 minutes. 5 cleaners work a total of 5 × 165 = 825 minutes.

If there are 3 workers then it will take 825 ÷ 3 = 275 minutes, which is 4 hours and 35 minutes each.

The same amount of work needs to be done regardless of how many people are doing it. If payment is made by the hour then the total cost should be the same. While that is not the same scenario that applies here, it is a useful starting point.

b 5 cleaners are paid for 3 hours.

3 cleaners are paid for 5 hours.

The total is the same.

7 **a** Total number of students = 720

$$\frac{20}{720} = 0.0278$$

Year 7 = 243 × 0.0278 = 6.76, which rounds off to 7 students.

Year 8 = 176 × 0.0278 = 4.89, which rounds off to 5 students.

Year 9 = 162 × 0.0278 = 4.5, which rounds off to 5 students.

Year 10 = 88 × 0.0278 = 2.45, which rounds off to 2 students.

Year 11 = 51 × 0.0278 = 1.42, which rounds off to 1 student.

b The stratified sampling method ensures that bigger year groups have more representatives.

Year 11 students are quite important in any school. Only having one Year 11 student on the council might not be a good idea, as that one might not share the same opinions as the majority. Year 7 students are new and might not have mature ideas about particular ideas/needs of older students. Hence for such a council it might be considered best to have 4 students from each year group.

> Take a stratified sample.

> There are lots of things you might write here. Think about the scenario and decide what would be sensible.

a Glass 1 contains $\frac{1}{4}$ units of squash.

Glass 2 contains $\frac{1}{6}$ units of squash.

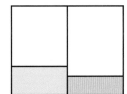

Pouring the two glasses together results in a total volume of 2 units.

The squash from glass 1 now represents $\frac{1}{8}$ of the combined drink.

The squash from glass 2 now represents $\frac{1}{12}$ of the combined drink.

$\frac{1}{8} + \frac{1}{12} = \frac{3}{24} + \frac{2}{24} = \frac{5}{24}$
So the resulting drink is $\frac{5}{24}$ squash.

b The strength of the final drink will be between the two strengths of the original drinks. $\frac{2}{5}$ and $\frac{3}{10}$ are both less than $\frac{1}{2}$ so it is not possible to make a drink that is exactly $\frac{1}{2}$ apple juice.

With questions like this it can help to assign a volume to the glass, be it a numerical value or an algebraic one.

Assume the volume of each glass is 1 unit so that you can just work with the fraction in the question.

Notice that this will produce a drink that is weaker than glass 1 and stronger than glass 2.

 a 10% of 90 g = 9 g

5% of 90 g $= \frac{9\,g}{2} = 4.5\,g$

The low-fat yogurt contains more sugar per pot (5.9 g compared to 4.5 g for the 'full-fat' version).

b $\frac{5.9\,g}{90\,g} = 0.06555...$

$0.06555... \times 100\% = 6.555...\%$

$\frac{4.5\,g}{90\,g} = 5\%$

The low-fat yogurt contains approximately 6.6% of the GDA of sugar.

The full-fat yogurt contains 5% of the GDA of sugar.

Calculate the amount of sugar in grams in the second yogurt.

Conclusion

First consider the low-fat yogurt.

Next consider the full-fat yogurt.

Conclusion

10 Grace needs to gain $\frac{5}{95}$ whereas Meg needs to lose $\frac{5}{105}$

$\frac{5}{95} = 5.263\%$

$\frac{5}{105} = 4.762\%$

The percentages are not the same even though the amounts of chocolate buttons involved are equal.

11 a $(x - 6) \times (x - 1)$

$= x^2 - 7x + 6$

So Valerie is correct.

Area involves multiplying, so it is likely that x^2 will be involved, so Valerie's expression looks more plausible. To work out the area of a rectangle find length × width.

b, c $x^2 - 7x + 6 - 2 = 2(2x - 7)$

$x^2 - 7x + 4 = 4x - 14$

$x^2 - 11x - 18 = 0$

$(x - 9)(x - 2) = 0$

so $x = 9$ or $x = 2$

If $x = 9$ then the dimensions of the rectangle are 3 by 8.

If $x = 2$ then one of the sides has negative length, which is impossible (which Doug found funny).

The expression for the perimeter is $2(2x - 7)$. Form an equation and solve for x.

12 a $\left(\frac{6}{20}\right)^2 + \left(\frac{5}{20}\right)^2 + \left(\frac{7}{20}\right)^2 + \left(\frac{2}{20}\right)^2 = \frac{114}{400} = \frac{57}{200}$

b $1 - \frac{57}{200} = \frac{143}{200}$

The easiest way to do this is to realise that picking a different colour the second time is like not picking the same colour.

c $\frac{6}{20} \times \frac{5}{19} + \frac{5}{20} \times \frac{4}{19} + \frac{7}{20} + \frac{6}{19} + \frac{2}{20} \times \frac{1}{19} = \frac{94}{380} = \frac{47}{190}$

d When he does not put the first marker back the probability that he gets two the same colour goes down.

This makes sense because if green was chosen first and then put back in, there is a higher chance green will be picked again than if there were fewer green pens available.

a £15000 = 20 m² × k

$$\frac{15000}{20} = k$$

k = £750 per square metre

$$a = \frac{23\,000}{750} = 30.666\,666 \text{ m}^2$$

But as the area must be an integer value, $a = 30\,\text{m}^2$

b 4 men would take 4 weeks, meaning there are 16 man-weeks of work.

To get the job done in 3 weeks there would need to be 16 ÷ 3 = 5.33333 men. So the builder could employ 5 labourers (making 6 men in total) to finish the job in less than 3 weeks.

c The total number of hours is 4 men × 4 weeks × 5 days per week × 9 hours per day = 720 man-hours of work.

With 6 workers that means each worker needs to do 720 ÷ 6 = 120 hours of work.

They work 9 hours per day, which means they need 120 ÷ 9 = $13\frac{1}{3}$ days.

This is two 5-day weeks, 3 days and 3 hours. If they start on a Monday they will do two full weeks and will finish on the following Thursday at 11 o'clock.

d The 3 hours that the 6 people would need to work on that final day is a total of 18 hours. The builder can do that by himself over two 9-hour days and still finish by the end of the three weeks (assuming there are no jobs that require two people to do them together, such as lifting heavy objects).

e 26 + 39 + 25 = 90 parts altogether, of which 25 parts are profit.

This means the profit is $\frac{25}{90}$ = 27.7777%

Job price includes nearly 28% profit.

Cost = area × constant of proportionality

Rearrange the equation to calculate the value of k.

You could work out how much the work costs, or you could just use the ratio that is given.

14 **a** Carpet Lay = £85, Underfoot = £105

b Underfoot

c $6\,m^2$

d Underfoot

e £125

These answers to parts **d** and **e** assume that the lines on the graph continue as straight lines.

15 **a**

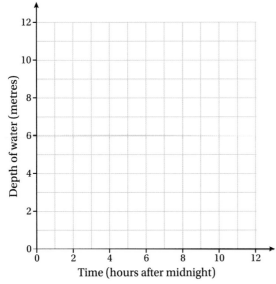

b Just after 6 am and just before 11 am (need to read off the graph as accurately as possible).

c Yes, in the second low tide period.

You know that the tide goes in and out roughly twice a day. Use this fact to help you answer the question.

16

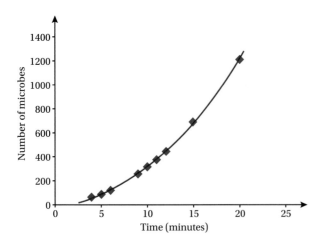

It is sensible to draw a graph.

This graph has a smooth slope, so it can be joined with a curve.

a Reading off the graph at 8 minutes there are about 200 microbes.

b 50 minutes is a long way away from the data, so the pattern might not continue (for example the microbes might get too hot and die off).

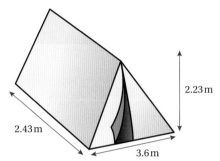

Peter is 2.03 m tall so the tent needs to be at least this long. You might assume that to be comfortable Peter will need the tent to be a little longer, for example an extra 20 cm at his head and at his feet.

To decide how wide the tent needs to be you have to make an assumption about the 'width' of Peter and his three friends. You want it to be comfortable so you should be generous with your estimate. Perhaps allowing 80 cm per person is reasonable.

Peter also needs to be able to stand up in the tent. If the tent is 2.03 m tall then Peter would just about be able to stand up in the middle (but his head would touch the top). So again, make it more comfortable and give him a little extra room, say 10 cm.

18 **a**

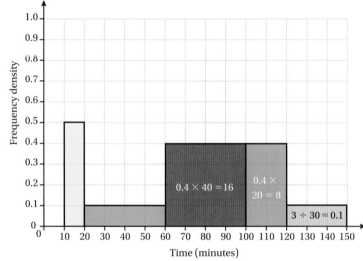

b The missing number in the table is 24.

c On the histogram, to the right of 100 mins there are 8 + 3 = 11 people. Because the data has been grouped you don't know for certain about the 8, but there are at least 3 students who play games for longer than 100 minutes (they are the ones in the 120–150 minute group).

Use the graph and some common sense to help you with this part of the question.

1

a $x = 4 + \dfrac{5}{x}$

> You could start with some algebra.

$x^2 = 4x + 5$

> Multiply by x.

$x^2 - 4x - 5 = 0$

$(x + 1)(x - 5) = 0$

> Factorise.

so $x = -1$ or $x = 5$

A positive number is required, so $x = 5$.

b $5 = 4 + \dfrac{5}{5}$

> Look back at part **a**.

Now $y = 5 + \dfrac{6}{y}$, and it is clear that $y = 6$ will work.

c This means that if $z = (n - 1) + \dfrac{n}{z}$ then $z = n$.

---

2 The red cross must be opposite the number 8.

> One way to answer this question is to draw a diagram showing the faces.

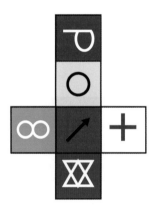

> Another is to spot that the cross appears on the same picture in the question as the P, the O, the arrow and the crossed triangles.

3

a 80% of £10 is £1 × 8 = £8. 75% of £12 is $\dfrac{3}{4}$ of £12 = £9. 75% of £12 is more money.

> There are lots of ways of carrying out these comparisons. Some of them are shown here.

b 50% of 1 is a half. 25% of 2 is also a half. They are the same.

c A 10% discount would be £1.299, which is less than the £1.99 postage charge, so a 5% discount is not enough and it would be cheaper to buy it in the shop.

> You don't need to do any difficult calculations.

4

Pattern number	1	2	3	4
Number of metal rods	6	12	18	24
Number of wooden rods	7	12	17	22

You could draw a few more diagrams and could then make a table of results to see whether that will help you see what is going on (although it won't give reasons for it).

a 18

The number of metal rods goes up by 6 each time because each new panel has 6 metal rods in it.

b 22

The number of wooden rods goes up by 5 each time because when a new panel is added, the left-hand side is joined to the previous panel.

c $6n$

d The $6 \times 12 = 72$ metal rods will cost £144, and the $5 \times 12 + 2 = 62$ wooden rods will cost £155.

The total cost is £299.

5 It could be:

a triangular prism

If you are only given the side elevation then there are lots of things it could be. All you know is that there is no 'crease' on that side and it looks like a triangle.

a triangular-based pyramid

a square-based pyramid

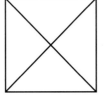

a cone

another shape!

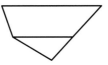

6 **a** The blue graph is MoveIt Builders.

> Read the information from the graph.

b MoveIt Builders = £15

c £14

> £70 for 5 days, daily rate = £70 ÷ 5

d Less. It is £12 daily, so £2 more per day.

> MoveIt Builders £75 for 5 days (75 – 15 = 60, 60 ÷ 5)

e £75

f £183

> £15 + (£12 × 14)

g MoveIt Builders

> JCEs would charge £14 × 14 = £196

h £13 cheaper

---

 7 **a**

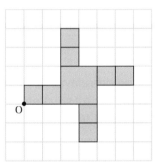

b 0 lines of symmetry

c Rotational symmetry order 4

d 1 Draw a circle of radius 1 cm.

> There are several ways this image could be created. Here is one possible set of instructions.

   2 Enlarge this circle by a scale factor of 2 about the centre of the circle.

   3 Translate the small circle 2 cm to the left.

   4 Translate the small central circle 2 cm to the right.

8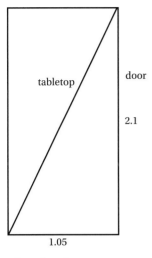

The tabletop diameter is 240 cm = 2.4 metres, which is more than the height of the door. To check whether it will fit when tilted at an angle, work out the length of the diagonal of the doorway.

$a^2 + b^2 = c^2$

$2.1^2 + 1.05^2 = c^2$

$5.5125 = c^2$

$c = \sqrt{5.5125} = 2.347...\,\text{m}$

Arthur's new table has a diameter of 240 cm. Even if Arthur turns the table to try to go through the door diagonally, it will not fit – it is more than 5 cm too big.

9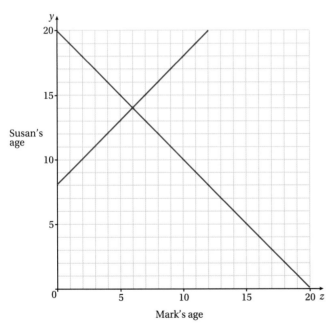

You could draw a graph as in this example solution. An alternative would be to solve the equations.

Their ages total 20, so $z + y = 20$.

Susan is 8 years older than Mark so $y = z + 8$.

The lines cross at $z = 6$ (Mark's age) and $y = 14$ (Susan's age).

10 a You know it is inversely proportional, so it must look like this:

If it is inversely proportional then one quantity goes up while the other goes down.

$$y = \frac{k}{x}$$

$$20\,000 = \frac{k}{4}$$

$$k = 80\,000$$

A possible formula is:

$$\text{number of bacteria} = \frac{80\,000}{\text{number of hours}}$$

b $\frac{80\,000}{8} = 10\,000$

c There will be problems when the time is very small. At zero hours it doesn't work at all.

---

11 $f(x) = (x - 2)(x + 2)$

You could answer this question algebraically.

$f(x) + 4 = (x - 2)(x + 2) + 4$

$\qquad = x^2 + 2x - 2x - 4 + 4$

$\qquad = x^2$

Alternatively you could draw a graph.

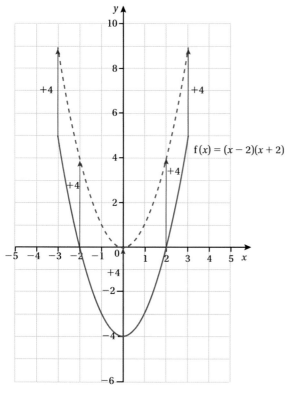

Both of these methods show that she is correct.

12

a $\frac{2}{3}$ as a decimal is 0.666... so $\frac{2}{3} > 0.66$.

It is much easier to compare them if they are both fractions or are both decimals.

b $0.25 = \frac{1}{4}$

$\frac{8}{32} = \frac{1}{4}$, so $0.25 = \frac{8}{32}$

In this pair, the fraction can actually be simplified. This will make the comparison easier.

c $\frac{3}{11} = 0.272727...$

so $\frac{3}{11} < 0.273$

Use short division to evaluate the decimal equivalent to $\frac{3}{11}$.

In this question it is easier to compare the decimals (finding a common denominator to compare the fractions is not as easy).

d $0.05 = \frac{5}{100} = \frac{1}{20}$,

$\frac{1}{20} > \frac{1}{22}$, so $0.05 > \frac{1}{22}$

e Here is a diagram showing 0.49.

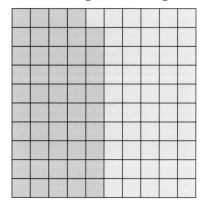

Now if you shade 0.4999 of the shape you get this (there is a tiny part of that top square that hasn't been shaded):

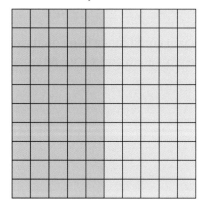

If you keep on doing this you get closer and closer to 0.5 each time, so we say that 0.4999 ... = 0.5.

a, b If h is the number of hamsters and c is the number of cages,

$h = 3c + 1$ and $h = 4(c - 1)$

Subtracting the first equation from the second, $c - 5 = 0$, so $c = 5$ and $h = 16$.

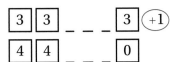

The three hamsters from the final cage, and the extra one, have been shared out and put into the other cages. That means there are now 4 cages with 4 hamsters each and one empty cage.

There are 5 cages and 16 hamsters.

> You don't know how many hamsters there are, or how many cages, which makes this difficult.

> You could write two equations and then solve them.

> An alternative would be to use a diagram.

---

 a If three consecutive even numbers add up to 228 then the middle one is $\frac{1}{3}$ of 228, the smallest one is 2 less than that and the biggest one is 2 more.

This gives 76 as the middle number so the answers are:

74, 76, 78

b Similar reasoning gives the middle number as $291 \div 3 = 97$.

The three numbers are 95, 97, 99.

c Half of 1301 is 650.5, so the two numbers must be either side of $\sqrt{605.5}$. This is 25.504... so the two numbers are 25 and 26.

d If the two numbers were equal then they would each be 40 (because their sum is 80). To get a difference of 6 you need to subtract 3 from one and add 3 to the other, giving 37 and 43.

> You could make a quadratic using $x^2 + (x + 1)^2 = 1301$ and solve it. An alternative would be to look for a rough average, as in this example.

> Alternatively you could make two equations and solve them (e.g. $x - y = 6$ and $x + y = 80$).

a If a to the power of something is 1 then the power equals zero.

$(a^x)^y = a^{xy}$, so either $x = 0$ or $y = 0$. The other one can be anything.

The only exception is if $a = 1$, when x and y can be anything.

b $(a^x)^y = a^1$ means that $x \times y = 1$.

> You may find that there are some other restrictions. For example, if a is zero or negative.

c $(a^x)^y = \sqrt{a}$ means that $x \times y = \dfrac{1}{2}$

---

 a $24.6 : 24$

$246 : 240$

$123 : 120$

> Start with the basic ratio, multiply by 10 to get integers, then divide by 2 to simplify.

b The ratios of times is $123 : 120$ whatever units are used.

> You could start again, having converted the times to minutes, but ratios don't have units.

c $6.4 \times 10^{23} : 6.0 \times 10^{24}$

$64 : 600$

$8 : 75$

d $27 : 9$

$3 : 1$

> The two mountains are approximately 27 km and 9 km tall.

17

> This involves several steps. Start with what you know: the square has area $4x^2 - 12x + 9$.

$4x^2 - 12x + 9 = (2x - 3)(2x - 3)$

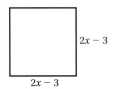

> Usually you would square root the area to work out the length of a side. Try factorising the expression.

Each side of the square is $2x - 3$. The perimeter is therefore $4(2x - 3)$ or $8x - 12$.

---

18 a

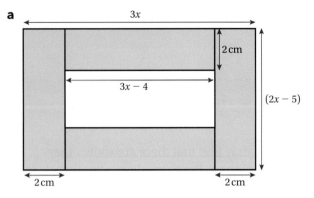

This is one way to split up the shape.

The top rectangle has area: $2(3x - 4) = 6x - 8$

The left rectangle has area: $2(2x - 5) = 4x - 10$

Altogether all four rectangles add up to $20x - 36$.

b $20x - 36 = 204$

$20x = 240$

$x = 12$

This means $3x$ is 36 and $(2x - 5)$ is 19.

The dimensions are 36 cm by 19 cm.

19 a The first number in each product is n and the second one is $3n - 1$.

The nth term is therefore $n(3n - 1)$ or $3n^2 - n$.

b 234 is divisible by 9, $234 \div 9 = 26$

Knowing the factors of 234 is helpful here.

$234 = 9 \times 26$

20 $150\,000\,\text{ml} \div 1000\,\text{ml} = 150$ bottles

Work out how many bottles of cola are required.

$150 \times £1.49 = £223.50$

Work out the total cost of the cola.

$150\,000 \div 400 = 375$ tins

Compare that with the number of tins of beans.

$375 \times 54\text{p} = £202.50$

Work out the total cost of the beans.

It will be cheaper to use the beans.

A quicker way would be to say that 1 ml of beans costs $54\text{p} \div 400 = 0.135\text{p/ml}$, whereas 1 ml of cola costs $149\text{p} \div 100 = 0.149\text{p/ml}$ so the beans are cheaper.

21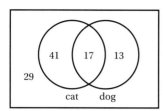

a $\dfrac{29}{200}$

b $\dfrac{41}{200}$

A Venn diagram will help to show what is going on.

Fill the 17 in first.

Then use the information about how many people own a dog.

Then use the information about how many people own either a dog or a cat but not both.

Finally, fill in the ones who own neither a dog nor a cat.

---

22 **a** $4x + 5y = 205$ (equation 1)

$3x + 7y = 235$ (equation 2)

b, c

$12x + 28y = 940$

$12x + 15y = 615$

$13y = 325$

$y = 25$

This means a banana costs 25 pence.

$4x + 5 \times 25 = 205$

$4x = 80$

$x = 20$

so an apple costs 20 pence.

Let x be the price of each apple in pence, and y be the price of each banana in pence.

One way to solve this is to get the same number of xs in each equation.

Multiply equation 2 by 4.

Multiply equation 1 by 3.

Subtract.

Divide by 13.

Substitute $y = 25$ into equation 1.

d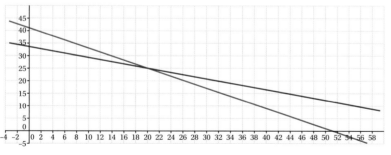

You can draw graphs of each of the lines and find the coordinates of the point where they cross.

23   **a**   There are two possibilities:

Rotate A 180° about (0, 3).

Enlarge A by a scale factor of –1 about point (0, 3).

b   There are a lot of possibilities:

Do two reflections (e.g. in the line $x = 0$ and then in the line $y = 3$).

Do a rotation and a translation (e.g. rotate A 180° about (–2, 3) and then translate it 4 units to the right.

Do a translation and then a rotation, or two enlargements, or an enlargement and a translation.

---

24   **a**

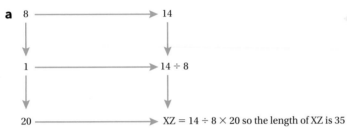

Draw a proportion diagram to work out XZ.

XZ = 35 cm

Triangle ABC has area 80 cm².

Triangle XYZ has area 245 cm².

b   The ratio is 80 : 245, which simplifies to 16 : 49.

c   16 and 49 are both square numbers.

The ratio of the sides is 4 : 7.

The ratio of the areas is $4^2 : 7^2$.

---

25 **a**

x	-2	-1	0	1	2	3	4	5	6
$f(x)$	11	2	-5	-10	-13	-14	-13	-10	-5

b The turning point is at $(3, -14)$.

Drawing a graph might help.

c $x^2 - 6x - 5 = 0$

$(x - 3)^2 - 9 - 5 = 0$

$(x - 3)^2 - 14 = 0$

$(x - 3)^2 = 14$

$x - 3 = \pm \sqrt{14}$

$x = 3 \pm \sqrt{14}$

The graph only gives approximate versions of where it crosses the x-axis. You could complete the square or use the formula to get these answers.

d The graph of $y = -f(x)$ shows that the turning point is at $(3, 14)$.

e This is the same as part **c**.

26 **a** $f(x - 8)$ moves the original function to the right by 8, whereas the vector moves it down by 8.

If this is not clear then you could draw graphs of $y = f(x)$ and $y = f(x - 8)$.

b He needs to move it back up by 8 and then to the right by 8, so the vector is $\begin{pmatrix} 8 \\ 8 \end{pmatrix}$.

27 **a**

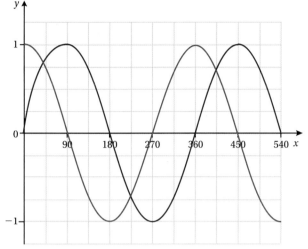

$y = \sin(x)$ is shown in red.

$y = \cos(x)$ is shown in blue.

b $f(x) = \cos(x - 90°)$

c $f(x) = \sin(x + 90°)$

d Translations.

---

28 **a** Each tile looks like this:

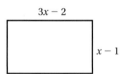

The area is $(3x - 2)(x - 1) = 3x^2 - 5x + 2$

Four of them have area $12x^2 - 20x + 8$

b The big square has sides of $4x - 3$, so the area is $(4x - 3)^2 = 16x^2 - 24x + 9$

c $4x^2 - 4x + 1$

To work out the inner square's area you can subtract part **a** from part **b**.

Or you could see on the diagram that the side of the inner square is $2x - 1$.

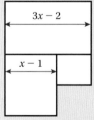

$(2x - 1)^2 = 4x^2 - 4x + 1$

a The area of the semicircle is
$\pi \times 25 \div 2 = 39.269...\,\text{cm}^2$.
This is not the same as Ollie's answer of $39.3\,\text{cm}^2$.

The height of the triangle $= \sqrt{75} = 8.660...\,\text{cm}$ – this is not the same as Ollie's answer of $8.7\,\text{cm}$.

Ollie has rounded too early in each calculation.

There are lots of things that might have gone wrong. You could work through Ollie's calculations and check the following:

Did he use the radius and not the diameter?

Did he halve the area of the circle?

Did he subtract when using Pythagoras' theorem?

Did he work out the area of the triangle incorrectly?

b $12.5\pi + 5\sqrt{75} = 82.571...\,\text{cm}^2$, which is $82.6\,\text{cm}^2$ to 1 dp.

30 **a**

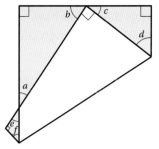

$b = 90 - a$

So the angles in the top left triangle are: a, 90, $90 - a$.

$a = e$

$f = 90 - a$

So the angles in the smallest triangle are:
a, 90, $90 - a$.

b, c and the right angle between them add up to 180°.

$b = 90 - a$, so $c = a$

$d = 90 - a$

The angles in the top right triangle are a, 90, $90 - a$.

The angles in the three triangles are identical so the triangles are similar.

b The square has sides of 8 cm so you know that
$CG + BG = 8$, $CE + ED = 8$, and AC and CB are both 4.

You can work out that $BG = 3$ and $CG = 5$ using Pythagoras:

$CG^2 = BG^2 + 4^2$

but $CG = 8 - BG$ so $(8 - BG)^2 = BG^2 + 16$

$$64 - 16BG + BG^2 = BG^2 + 16$$

$$48 = 16BG$$

$$BG = 3$$

You know from part **a** that all the triangles are similar, so $\dfrac{CE}{CG} = \dfrac{AC}{BG}$

$CE = \dfrac{4}{3} \times 5 = \dfrac{20}{3} = 6\dfrac{2}{3}$,

and $\dfrac{AE}{BC} = \dfrac{AC}{BG}$

$AE = \dfrac{4}{3} \times 4 = \dfrac{16}{3} = 5\dfrac{1}{3}$

$DE = 8 - 6 = 1$

$\dfrac{AE}{DE} = 5\dfrac{1}{3} \div 1\dfrac{1}{3} = \dfrac{16}{3} \div \dfrac{4}{3} = 4$, so the sides of the largest triangle are 4 times those of the smallest one.

For the triangles to be similar they must have the same angles. The paper is square so the corners are all 90°.

Start by looking at the top left triangle, and then work out the angles in all the triangles in relation to this.

Angles in a triangle add up to 180°.

Vertically opposite angles are equal.

Angles in a triangle add up to 180°.

Angles on a straight line.

Angles in a triangle add up to 180°.

Move from thinking about each right-angled triangle to thinking about the square.

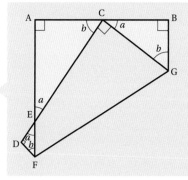

31 **a** **i** $\tan 60° = \dfrac{\text{opp}}{\text{adj}} = \dfrac{\sqrt{3}}{2} \div \dfrac{1}{2} = \sqrt{3}$

ii $\sin 60° = \dfrac{\text{opp}}{\text{hyp}} = \dfrac{\sqrt{3}}{2} \div 1 = \sqrt{3}$

iii $\cos 60° = \dfrac{\text{adj}}{\text{hyp}} = \dfrac{1}{2} \div 1 = \dfrac{1}{2}$

Work out the other lengths on the diagram.

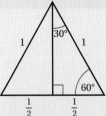

The height of the triangle can be worked out using Pythagoras' theorem:

$$\sqrt{1^2 - \left(\tfrac{1}{2}\right)^2} = \sqrt{\left(\tfrac{3}{4}\right)} = \dfrac{\sqrt{3}}{2}$$

b   To work out $\sin 60°$ you need $\dfrac{\sqrt{3}}{2} \div 1 \quad \left(\dfrac{\text{opp}}{\text{hyp}}\right)$

To work out $\cos 30°$ you need $\dfrac{\sqrt{3}}{2} \div 1 \quad \left(\dfrac{\text{adj}}{\text{hyp}}\right)$

The opposite side needed to work out $\sin 60°$ is the adjacent side needed to work out $\cos 30°$.

The same argument holds for $\sin 30°$ and $\cos 60°$.

1 a

+	1	2	3	4	5	6	7	8	9
1									
2									
3									
4									
5									
6									
7									
8									
9									

Systematically pair up every integer with every other integer.

The table illustrates all the possible calculations that can be made.

Notice that these are different calculations. They do not necessarily represent different answers.

The answer is 45.

b

	–9	–8	–7	–6	–5	–4	–3	–2	–1
–9	–18	–17	–16	–15	–14	–13	–12	–11	–10
–8	–17	–16	–15	–14	–13	–12	–11	–10	–9
–7	–16	–15	–14	–13	–12	–11	–10	–9	–8
–6	–15	–14	–13	–12	–11	–10	–9	–8	–7
–5	–14	–13	–12	–11	–10	–9	–8	–7	–6
–4	–13	–12	–11	–10	–9	–8	–7	–6	–5
–3	–12	–11	–10	–9	–8	–7	–6	–5	–4
–2	–11	–10	–9	–8	–7	–6	–5	–4	–3
–1	–10	–9	–8	–7	–6	–5	–4	–3	–2

The shaded cells show the 17 different answers you can get.

2

+ – ×: $5 + 2 - 10 \times -3 = 5 + 2 - -30 = 5 + 2 + 30 = 37$

+ × –: $5 + 2 \times 10 - -3 = 5 + 20 + 3 = 28$

– + ×: $5 - 2 + 10 \times -3 = 5 - 2 - 30 = -27$

– × +: $5 - 2 \times 10 + -3 = 5 - 20 - 3 = -18$

× + –: $5 \times 2 + 10 - -3 = 10 + 10 + 3 = 23$

× – +: $5 \times 2 - 10 + -3 = 10 - 10 - 3 = -3$

You can achieve a negative result in three ways.

Make a systematic list of the arrangements of the three operations and then calculate the answers.

3   **a** The median is 3, and because the mode of 10 numbers is 3 and there are only three different numbers, there must be at least four 3s.

_ _ _ _ 3 3 _ _ _ _
3 3 3 3 3 3 _ _ _ 13
The other numbers must be 8, 8, 13.

Start by working through the information you are given in order.

The mean is 6, so the total is 60. The range is 10, so if the lowest number is 3 you get this.

If the lowest is 2 then the highest is 12 and there are no combinations of 2, 3 and 12 that fit the rules.

1 3 3 3 3 3 11 11 11 11 also works.

If the lowest is 1 then the highest is 11.

b 3 3 3 71 and

1 3 3 73 will work

The mode and the median are both 3 so the middle two numbers are both 3. The mean is 20 so they all add up to 80.

c 3 3 7 67

The mode is 3 so there must be at least two 3s. The median is 5, so you get this.

d 3 3 17 57

e 3 3 27 47

f As long as two of the numbers are 3 and the other two are both odd and add up to 74 then this will work.

---

4 $\frac{2}{3} + \frac{1}{5} = \frac{10}{15} + \frac{3}{15} = \frac{13}{15}$

Here is a systematic way to do this.

$\frac{2}{3} - \frac{1}{5} = \frac{7}{15}$

$\frac{1}{5} - \frac{2}{3} = -\frac{7}{15}$

$\frac{2}{3} \times \frac{1}{5} = \frac{2}{15}$

$\frac{2}{3} \div \frac{1}{5} = \frac{2}{3} \times \frac{5}{1} = \frac{10}{3}$

$\frac{1}{5} \div \frac{2}{3} = \frac{1}{5} \times \frac{3}{2} = \frac{3}{10}$

a The biggest answer is $\frac{10}{3}$

b The smallest answer is $-\frac{7}{15}$

5 **a** If a and b are not negative and have to be whole numbers then either $a = 1$ and $b = 0$ or $a = 0$ and $b = 1$.

b This time they are positive, so 0 cannot be used.

> Make a systematic list.

1: $0.1 + 0.9 = 1$

2: $0.2 + 0.8 = 1$

3: $0.3 + 0.7 = 1$

⋮

9: $0.9 + 0.1 = 1$

There are 9 of them.

c 1: $0.01 + 0.99 = 1$

2: $0.02 + 0.98 = 1$

3: $0.03 + 0.97 = 1$

⋮

99: $0.99 + 0.01 = 1$

There are 99 of them.

d If a and b are allowed to be negative then $a + b = 1$ with the only restriction being that they are both integers. Whatever you choose as the value for a you can choose a value for b that works, because b is $1 - a$.

6 **a** **i**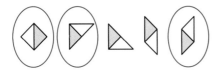

With two sandwiches it is possible to create three different polygons: a right-angled isosceles triangle, a parallelogram and a square.

> If you have two sandwiches (two triangles) then it might make sense to keep one of them fixed and move the other around it. Here the blue one is fixed and the white one moves.

> There are five different positions in which the second sandwich could be placed, three of which make different polygons.

ii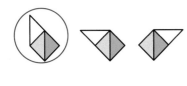

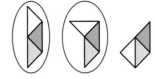

> If you have three sandwiches (three triangles) then it makes sense to start with the three arrangements for two sandwiches and move the third sandwich around these.

> The second and third shapes that start with a square are congruent to the first one – and the others that start with a square will also all be congruent or reflections.

> The third shape is congruent to the ones above. All the others that start with the parallelogram will be congruent to others that we already have.

> This is the only new shape formed by starting with the right-angled isosceles triangle.

With three sandwiches it is possible to create four different polygons: two different trapeziums and two different irregular pentagons.

iii

With four sandwiches it is possible to create nine different polygons: a rectangle, an irregular pentagon, three different irregular hexagons, an isosceles trapezium, a parallelogram, an isosceles triangle and a square.

b Yes.

Conclusion

If you have four sandwiches (four triangles) then it makes sense to start with the four arrangements for three sandwiches and move the fourth sandwich around these.

Again, some of these are identical or are reflections of each other.

These are the ones that are distinct.

This is almost a trick question. A square is an example of all the special quadrilaterals (it is also a rectangle, a parallelogram, a rhombus, a kite and a trapezium), so the original sandwich works for all of them.

---

 a Box 1: four cylinders of radius 10 cm and height 15 cm

Box 2: one cylinder of radius 20 cm and height 15 cm

Box 3: sixteen cylinders of radius 5 cm and height 15 cm

c Box 1: vol. of cylinders = $\pi \times 10^2 \times 15 \times 4 = 6000\pi$

Box 2: vol. of cylinder = $\pi \times 20^2 \times 15 = 6000\pi$

Box 3: vol. of cylinders = $\pi \times 5^2 \times 15 \times 16 = 6000\pi$

Because the total volume of the cylinders is the same each time, the unused space is the same in each box.

Did this answer surprise you? Or did you guess that all three boxes had the same amount of unused space?

If you look at each calculation for the volume of the cylinders, you will notice that each has $\pi \times 15$. The rest of the calculation for each volume multiplies to give the same answer, 400.

$4 \times 10^2 = 20^2 = 16 \times 5^2 = 400$

8 **a**

First card	Second card				
	J_1	J_2	Q	K	A
J_1		J_1J_2	J_1Q	J_1K	J_1A
J_2			J_2Q	J_2K	J_2A
Q				QK	QA
K					KA
A					

> There are two jacks and because they are different suits it is useful to call them J_1 and J_2.

b Of these 10 possibilities 1 involves the two jacks: $\frac{1}{10}$.

c From the table it is $\frac{3}{10}$.

---

9 **a** This table shows there are 78 pairs of numbers.

	1	2	3	4	5	6	7	8	9	10	11	12
1	1	1	1	1	1	1	1	1	1	1	1	1
2		2	1	2	1	2	1	2	1	2	1	2
3			3	1	1	3	1	1	3	1	1	3
4				4	1	2	1	4	1	2	1	4
5					5	1	1	1	1	5	1	1
6						6	1	2	3	2	1	6
7							7	1	1	1	1	1
8								8	1	2	1	4
9									9	1	1	3
10										10	1	2
11											11	1
12												12

> This table, and the way it is shaded, is a lovely way to show what is going on here.

The number in each cell is the HCF of the pair of numbers. The co-prime numbers are shown with a 1.

b There are 46 orange cells, so $\frac{46}{78}$ of the pairs are co-prime. This simplifies to give $\frac{23}{39}$.

10 **a** The number 1 is used in four lines. So is the number 2.

All the numbers from 1 to 7 are used four times so the total is $(1 + 2 + 3... + 7) \times 4 = 28 \times 4 = 112$.

As an alternative method, you could add up the totals for each line.

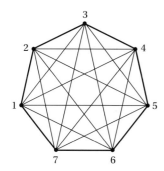

b In a hexagon each number joins to three others, so the total is $(1 + 2 + 3... + 6) \times 3 = 21 \times 3 = 63$.

In a pentagon each number joins to two others, so the total is $(1 + 2 + 3 + 4 + 5) \times 2 = 15 \times 2 = 30$.

The total of $63 + 30 = 93$ is less than 112.

11 You know the triangles are equilateral because the sides are all radii of the circles. This helps to work out the angle of 240°.

There are many ways to work this out. This method uses two sectors of circles (shown unshaded) and two equilateral triangles.

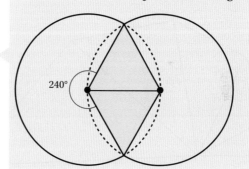

The perimeter is therefore: $\frac{240}{360} \times \pi \times d \times 2 = \frac{16}{3}\pi$.

The area of each sector is $\frac{240}{360} \times \pi r^2 = \frac{8}{3}\pi$.

The area of one triangle is $\frac{1}{2}ab\sin C = \frac{1}{2} \times 2 \times 2 \times \sin 60°$.

The total area is therefore $\frac{16}{3}\pi + 2\sqrt{3}$.

The cost of the pool is $\frac{16\pi}{3} \times £225 + \left(\frac{16\pi}{3} + 2\sqrt{3}\right) £800$
$= £19\,945$.

You know that all powers of 1 are 1, so this means that $2^b + 3^c = 244$.

This question can be worked out in several different systematic ways. Here is one.

You don't need to worry about negative powers of 2 and 3 because they will give fractions.

You could make a list of powers of 2 and powers of 3, but consider for a moment that 3^c is always odd. The only way to get an even result (like 244) is for 2^b to be odd.

So this must mean $2^0 = 1$ and $3^c = 243$.

3^0	1
3^1	3
3^2	9
3^3	27
3^4	81
3^5	243

This means that:

a can be anything

$b = 0$

$c = 5$

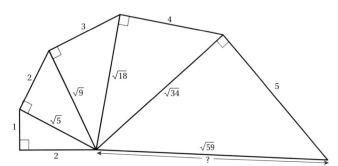

Use Pythagoras' theorem in the smallest triangle, but leave your answer as a surd because you will need to square it again to work out the hypotenuse of the second triangle, and so on.

The final hypotenuse is the square root of $2^2 + 1^2 + 2^2 + 3^2 + 4^2 + 5^2$.

This is $\sqrt{59}$.

14

Week number	1	2	3	4	5	6
Number of people at end of week	4	16	64	256	1024	4096

Molly gives the mixture away at the end of each week, so this table shows the number of people who have the mixture at the end of the week. Because each person shares theirs with 3 others it is multiplied by 4 each week.

a 256 people will have it after 4 weeks.

b 4096 people after 6 weeks.

c $4^{13} = 67\,108\,864$

So 13 weeks are required.

Find which power of 4 will give a number greater than 66 million.

d Not everyone will pass it on or keep their own one going. Babies and children would find it difficult. After a while everyone you know will have it and you won't be able to find anyone new to pass it on to.

 15

N	Fraction	Decimal	Recurring or terminating?
1	$\frac{1}{1}$	1	Terminating
2	$\frac{1}{2}$	0.5	Terminating
3	$\frac{1}{3}$	0.3333...	Recurring
4	$\frac{1}{4}$	0.25	Terminating
5	$\frac{1}{5}$	0.2	Terminating
6	$\frac{1}{6}$	0.1666...	Recurring
7	$\frac{1}{7}$	0.142857142857...	Recurring
8	$\frac{1}{8}$	0.125	Terminating
9	$\frac{1}{9}$	0.1111...	Recurring
10	$\frac{1}{10}$	0.1	Terminating

Four of the fractions , where N is any positive integer from 1 to 10, are recurring.

If the numerator is 1 then the terminating ones must have denominators whose only prime factors are 2 or 5.

1 **a** Jyoti's tower = $1 + 2 + 3 + 4 + 5 + 6 = 21$ cm

Indira's tower = $7 + 8 + 9 + 10 + 11 + 12 = 57$ cm

Indira's tower is 36 cm taller.

> One way to do this is to write out the heights of the two towers.

b $57 + 21 = 78$ cm

One possible answer is $1 + 2 + 3 + 10 + 11 + 12 = 39$

and $4 + 5 + 6 + 7 + 8 + 9 = 39$

> The total height of all the cubes is 78 cm.
>
> $78 \div 2 = 39$ cm gives the height of each tower when the two towers are the same. Each girl must have a combination that is 39 cm high.

2 There are eight numbers.

The biggest is 16.

The range is 15, so the smallest is $16 \quad 15 = 1$.

The mean is 7.5 so they all add up to $7.5 \times 8 = 60$.

The mode is 3 and 5 so there are the same number of 3s and 5s, and there must be two or three of each.

The six numbers that must be included are 1, 3, 3, 5, 5, 16 and the other two numbers must be between 1 and 16. The six numbers 1, 3, 3, 5, 5, 16 add to 33, so the other two numbers add to 27.

The possible solutions are therefore:

1, 3, 3, 5, 5, 12, 15, 16

1, 3, 3, 5, 5, 13, 14, 16

> What does each piece of information tell you?
>
> Write down what you know.

3 **a** The mode is 2 and the median is 5 so there must be:

2 2 5 _ _

The range is 15 so the biggest number is 17, and the mean is 7 so they all need to total 35, giving 9 as the other number:

2, 2, 5, 9, 17

> Write down the things you know and then see what you can work out.

b For the range, mode and median from before there has to be 2 2 5 _ 17

These total 26. The mean is an integer so the total must be a multiple of 5. The biggest number that is allowed in the space is 16 but this gives 42 altogether so it must be 14 and the mean is 8.

c This time the mean is not stated to be an integer.

2, 2, 5, 6, 17 gives the smallest mean of $32 \div 5 = 6.4$

4　**i**　False

　　ii　True

If you add the digits of the number together you get 48, which is divisible by 3, so the number is also divisible by 3.

　　iii　False

The number is a multiple of 3, so it cannot be prime. It is also even.

　　iv　True

The number 12 345 678 910 is in the 5 times table, so if you divide 12 345 678 912 by 5 then the remainder is 2.

　　v　True

The last two digits are 12, which is a multiple of 4. If a number is a multiple of 4 then the last two digits will also be a multiple of 4. This means the number is a multiple of 4.

5　Grandad's age is $2x$, Geoff's age is x and Paul's age is $x - 29$.

One way to do this is to call Geoff's age x and to work out the other ages in terms of x.

In total this is $2x + x + x - 29$, which the question says equals 131.

$4x - 29 = 131$, so $4x = 160$ and $x = 40$.

6　Call the daughter's age x. The father is then $5x$ and the mother is $5x - 9$.

You could start by using some algebra.

Adding these up gives $11x - 9$, which is equal to 134 years.

Solve $11x - 9 = 134$ to get $x = 13$.

The mother is therefore $5 \times 13 - 9 = 56$ years old.

7 **a** $1:8$ — Ratio of teachers to students.

$x:140$ — Let number of teachers = x.

$140 \div 8 = 17.5$ — Number of teachers must be a whole number.

18 teachers

b $21 \times 8 = 168$ — Every adult can have 8 students to supervise.

$168 - 140$ — A total of 168 students can go.

28 more students

c The ratio of students to teachers is

$140:21$

$20:3$

d 21 teachers + 140 students = 161 seats needed — First find the number of coaches needed.

$4 \times 42 = 168$ seats

So 4 coaches are needed.

4 coaches need 4 drivers.

Total adults = 25.

Students : Adults

$140:25$

$28:5$

e $\dfrac{25}{165} = \dfrac{5}{33}$ — Total adults = 25.

Total in the group = 25 + 140 = 165.

8

A clear, well-labelled diagram will be useful. This one focuses on the important triangle.

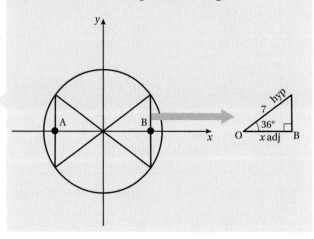

OB = cos 36° × 7 cm = 5.663... cm

To work out x you need to use trigonometry.

B: (5.66, 0)

Now interpret this in terms of the coordinates of point B (values rounded to 2 dp).

A: (–5.66, 0)

Point A is a reflection of point B in the y-axis.

9 **a**

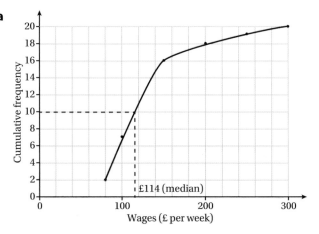

The usual way to work out the median from a grouped frequency table is to draw a cumulative frequency graph.

Weekly earnings (£)	Frequency	Cumulative frequency
61–80	2	2
81–100	5	7
101–150	9	16
151–200	2	18
201–250	1	19
251–300	1	20

This gives a median of about £114, which is not helpful to the manager as it is below £130.

b The median won't increase.

The mean salary is (7 × 90.5 + 9 × 125.5 + 2 × 175.5 + 1 × 225.5 + 1 × 275.5) ÷ 20.

This is £130.75, so you could say that the average is now above the national average.

You might be interested to see what happens to the median when worked out in this way!

10

0.74 or $\dfrac{74}{100}$	0.46 or $\dfrac{46}{100} = \dfrac{23}{50}$	0.9 or $\dfrac{90}{100} = \dfrac{9}{10}$
0.86 or $\dfrac{129}{150} = \dfrac{43}{50} = \dfrac{86}{100}$	0.7 or $\dfrac{70}{100} = \dfrac{7}{10}$	0.54 or $\dfrac{54}{100} = \dfrac{27}{50}$
0.5 or $\dfrac{50}{100} = \dfrac{1}{2}$	0.94 or $\dfrac{94}{100} = \dfrac{47}{50}$	0.66 or $\dfrac{33}{50} = \dfrac{66}{100}$

Remember that in a magic square all the rows and columns and both main diagonals add up to the magic number. The key idea here is to convert everything into the same type of number. It is easiest to work either in decimals or in fractions with a denominator of 100. The diagram shows possible answers: you only need one answer in each square.

 11

a e.g. $3 = \dfrac{2}{1} + 1$, or $3 = \dfrac{3}{1} + 0$, etc.

b e.g. $3 = \dfrac{4}{2} + 1$, or $3 = \dfrac{8}{8} + 2$, etc.

Since c must be an integer, the fraction must be an integer, too.

This means that a must be a multiple of b.

c e.g. $3 = \dfrac{-2}{-1} + 1$, or $3 = \dfrac{6}{-3} + 5$, etc.

The fraction will be positive if you use negative values for both a and b.

 12

1 pint (imperial) is 0.568 litres (568 ml).

9 pints = 568 ml × 9 = 5112 ml

We know there are two different units in the question: millilitres and pints.

Here we have converted pints into millilitres.

5112 ml = approximately 5000 ml

$= 5 \times 10^3$ ml of blood in James's body

The question asks you to work with approximate figures.

There are 5×10^9 red blood cells per millilitre of blood, so in total James has about
$5 \times 10^9 \times 5 \times 10^3 = 25 \times 10^{12}$
$= 2.5 \times 10^{13}$ red blood cells.

10% of $2.5 \times 10^{13} = 0.25 \times 10^{13}$

The number of remaining blood cells is
$2.5 \times 10^{13} - 0.25 \times 10^{13} = 2.25 \times 10^{13}$.

13 There are 4 aces. The probability of picking 2 aces is:

$$\frac{4}{52} \times \frac{3}{51} = \frac{12}{2652}$$

There are two red kings, so the probability of picking 2 red kings is:

$$\frac{2}{52} \times \frac{1}{51} = \frac{2}{2652}$$

Picking two aces is 6 times as likely.

> You could start by writing down the probability of picking one ace. Then, if you are lucky and have picked an ace, write down the probability of picking an ace when there are only 51 cards left.

14

		Socks						
		G	**G**	**B**	**B**	**B**	**BLU**	**STR**
Ties	**P**	a	a					
	BLU						d	
	PUR							b
	G	c d	c d					

> In probability questions a diagram of some kind is often helpful. Here the letters in the body of the table refer to each part of the answer.

a $\frac{2}{28} = \frac{1}{14}$

b $\frac{1}{28}$

c $\frac{2}{28} = \frac{1}{14}$

d $\frac{3}{28}$

e $1 - \frac{3}{28} = \frac{25}{28}$

15

+	4	5	6	7	8
4			10		
5		10			
6	10				
7					
8					

p(4 kits and 6 kits) = 0.1 × 0.3 = 0.03

p(5 kits and 5 kits) = 0.2 × 0.2 = 0.04

p(6 kits and 4 kits) = 0.3 × 0.1 = 0.03

Any of these outcomes are possible, so add them together to get 0.1.

The probabilities should add up to 1.

A table might help but note that this shows the different possibilities and not the probabilities (because they are not all equally likely).

---

 16   **a** Let x be the number of glasses of lemonade and y be the number of brownies.

$30x + 20y \leqslant 350$

b If n is the number of friends, then $50n \leqslant 350$

so $n \leqslant 7$.

Each glass of lemonade needs 30 g sugar, so altogether the drinks needs $30x$ sugar.

Each brownie needs 20 g sugar, so altogether the snacks need $20y$ sugar.

In total this must be less than or equal to 350.

If everyone has one of each, then that is 50 g of sugar per person.

17

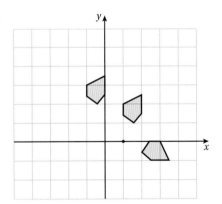

It seems sensible to work backwards through the transformations.

The question isn't clear as to whether the second, intermediate shape is drawn, so you might not have the shape starting from coordinate point (1, 2)

18 a

$$\sqrt{1^2 + 1^2} = \sqrt{2}$$

So the hypotenuse is $\sqrt{2}$.

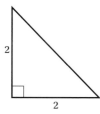

$$\sqrt{2^2 + 2^2} = \sqrt{8}$$

This shows that the hypotenuse is $\sqrt{8}$.

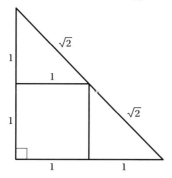

So $2\sqrt{2} = \sqrt{8}$

It seems sensible to work out the length of each hypotenuse (because square roots are involved).

b

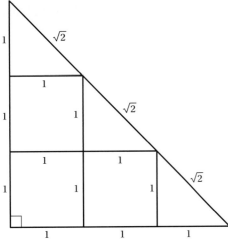

$3 \times \sqrt{2}$ looks like a triangle with sides 3 times those of the small triangle.

The big triangle above is identical to the triangle below.

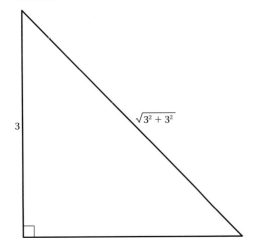

This shows that $\sqrt{3^2 + 3^2} = \sqrt{18}$

19

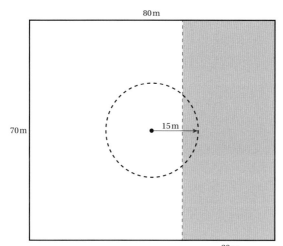

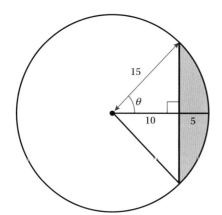

This is not easy to visualise, so a sketch is a good idea.

You need to know the shaded area that the goat can reach.

$\cos\theta = \dfrac{10}{15}$

so $\theta = 48.189\ldots$

The area of the sector is $\dfrac{2\theta}{360} \times \pi \times 15^2$

The area of the triangle is

$\dfrac{1}{2}ab\sin C = \dfrac{1}{2} \times 15 \times 15 \times \sin(2\theta)$

This gives the shaded, goat-eaten area as
$189.24\ldots - 111.80\ldots = 77.437\ldots$

The cost of planting the wheat is $30 \times £65 = £1950$

The takings are $£4.50 \times (30 \times 70 - 77.437\ldots) = £9101.53$

The farmer's profit is $£9101.53 - £1950 = £7151.53$

20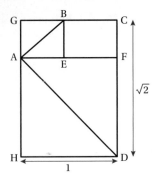

The question states that it is a kite, so AD = CD and AB = BC.

AD and CD are both $\sqrt{2}$.

AH = 1

> Because it is folded across the width of the rectangle, which is 1.

so GA = $\sqrt{2}$ – 1

GA is folded to meet AE so AE = ($\sqrt{2}$ – 1) and EF = 1 – ($\sqrt{2}$ – 1) = 2 – $\sqrt{2}$

BC is the same as EF, so sides AB and BC are both 2 – $\sqrt{2}$

The four sides total: $\sqrt{2} + \sqrt{2} + 2 - \sqrt{2} + 2 - \sqrt{2} = 4$

> This question can be answered in lots of ways. Here is one that does not involve the use of Pythagoras' theorem.

21

$KE = \frac{1}{2}mv^2$, where m is the mass in kg and v is the speed in m/s.

The tennis ball has mass 56 g and velocity 120 miles per hour.

56 g = 0.056 kg so m = 0.056

$$120 \text{ miles per hour} = 120 \times \frac{8}{5} \text{km per hour}$$

$$= 192 \text{ km per hour}$$

$$= \frac{192\,000}{3600} \text{m/s} = 53.\dot{3}\,\text{m/s}$$

$$KE = \frac{1}{2} \times 0.056 \times (53.\dot{3})^2$$

$$= 79.6 \text{ joules}$$

Student 1 has not converted the units so that the mass is in kg and the speed is in m/s.

Student 2 appears to have worked out $(\frac{1}{2}mv)^2$ rather than $\frac{1}{2}mv^2$ (where only the v should be squared).

> The first thing you might want to do is to try to work this out for yourself.

CAMBRIDGE

Brighter Thinking

GCSE
MATHEMATICS
online

The complete online resource for GCSE Mathematics

GCSE Mathematics Online for AQA is our brand new interactive teaching and learning subscription service. This tablet-friendly resource supports both independent learning and whole class teaching through a suite of flexible resources that includes lessons, tasks, questions, quizzes, widgets and games.

▷ Allows teachers to set tasks and tests, auto-mark and compile reports to review student performance.

▷ A test generator so teachers can compile their own assessments.

▷ Interactive widgets to visually demonstrate concepts.

▷ Worksheets offering practical activities, discussion points, investigations, games and further practice.

▷ Walkthroughs that take students through a question step-by-step, with feedback.

▷ Quick-fire quizzes with leaderboards, providing an opportunity for question practice.

▷ Levelled questions that assess understanding of each topic.

▷ Covers both Foundation and Higher, offering flexibility for moving students between tiers.

▷ Resources organised into chapters corresponding to the Student Books, with explanatory notes for all topics.

▷ Contains material for all types of classroom set-up, including interactive whiteboards and projectors.

www.gcsemaths.cambridge.org

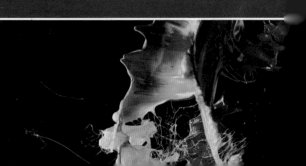